D0841040

McGraw-Hill Education

500 Review Questions for the MCAT: General Chemistry

Also in McGraw-Hill's 500 Questions Series

McGraw-Hill Education

500 Review Questions for the MCAT: General Chemistry

John T. Moore, EdD
Richard H. Langley, PhD

HALF HOLLOW HILLS
COMMUNITY LIBRARY
55 Vanderbilt Parkway
Dix Hills, NY 11746

New York   Chicago   San Francisco   Athens   London   Madrid
Mexico City   Milan   New Delhi   Singapore   Sydney   Toronto

Copyright © 2015, 2012 by The McGraw-Hill Education, Inc. All rights reserved. Printed in the United States of America. Except as permitted under the United States Copyright Act of 1976, no part of this publication may be reproduced or distributed in any form or by any means, or stored in a database or retrieval system, without the prior written permission of the publisher.

1 2 3 4 5 6 7 8 9 10   QFR/QFR   1 2 1 0 9 8 7 6 5

ISBN   978-0-07-183616-6
MHID   0-07-183616-0

e-ISBN   978-0-07-183617-3
e-MHID   0-07-183617-9

MCAT is a registered trademark of the Association of American Medical Colleges, which was not involved in the production of, and does not endorse, this product.

McGraw-Hill Education products are available at special quantity discounts to use as premiums and sales promotions or for use in corporate training programs. To contact a representative, please e-mail us at bulksales@mcgraw-hill.com.

CONTENTS

ABOUT THE AUTHORS

John T. Moore, EdD, has taught chemistry at Stephen F. Austin State University for more than 40 years, where he is the Director of the Teaching Excellence Center and Codirector of the Science, Technology, Engineering and Mathematics Center.

Richard H. Langley, PhD, has taught chemistry at the university level for over 30 years and has coauthored numerous books on the subject. He has written questions for the AP Chemistry exam and taught General Chemistry and Organic Chemistry MCAT review courses.

INTRODUCTION

Congratulations! You've taken a big step toward MCAT success by purchasing *500 Review Questions for the MCAT: General Chemistry.* We are here to help you take the next step and score high on the MCAT so you can get into the medical school of your choice.

This book gives you 500 multiple-choice questions that cover all the most essential course material. Each question is clearly explained in the answer key. The questions will give you valuable independent practice to supplement your regular textbook and the ground you have already covered in your classes.

This book and the others in the series were written by expert teachers who know the MCAT inside and out and can identify crucial information as well as the kinds of questions that are most likely to appear on the exam.

You might be the kind of student who needs to study extra a few weeks before the exam for a final review. Or you might be the kind of student who puts off preparing until the last minute before the exam. No matter what your preparation style, you will benefit from reviewing these 500 questions, which parallel the content and degree of difficulty of the questions on the actual MCAT. These questions and the explanations in the answer key are the ideal last-minute study tool for those final weeks before the test.

If you practice with all the questions and answers in this book, we are certain you will build the skills and confidence needed to excel on the MCAT. Good luck!

—Editors of McGraw-Hill Education

CHAPTER 1

The Basics

Questions 1–6 refer to the following passage.

A chemist can extract much information from the balanced chemical equation. The coefficients in the balanced chemical equation represent the number of particles, but also the number of moles, which can also be related to the mass of reactants consumed and the mass of product formed. This mass relationship between the reactants and the products is the reaction stoichiometry. Stoichiometry is the quantitative relationship as defined by the balanced chemical equation. This quantitative relationship can be expressed in either moles or mass or particles.

Normally one reactant is totally used up during the reaction. There will be leftover amounts of the other reactants. The reactant that is totally consumed during the reaction, the one that limits the amount of product(s) formed, is the **limiting reactant (LR)**.

1. Copper(II) nitrate decomposes when heated according to the following equation:

$$2 \, Cu(NO_3)_2 \rightarrow 2 \, CuO + 4 \, NO_2 + O_2$$

How many grams of NO_2 form when 2.0 moles of copper(II) nitrate decompose?
   (A) 46 g
   (B) 92 g
   (C) 180 g
   (D) 220 g

2. The following equation illustrates the formation of heme from ferrous ion and protoporphyrin IX:

$$C_{31}H_{28}N_4O_4(aq) + Fe^{2+}(aq) \rightarrow FeC_{31}H_{26}N_4O_4(aq) + 2\ H^+(aq)$$

What type of reaction is the formation of heme?

(A) Lewis acid-base
(B) Oxidation-reduction
(C) Single displacement
(D) Neutralization

3. The following equation illustrates the formation of heme from ferrous ion and protoporphyrin IX:

$$C_{31}H_{28}N_4O_4(aq) + Fe^{2+}(aq) \rightarrow FeC_{31}H_{26}N_4O_4(aq) + 2\ H^+(aq)$$

If the final pH of the solution was 5, what was the original pH of the solution?

(A) Equal to 5
(B) Less than 5
(C) Greater than 5
(D) Cannot be determined

4. The formation of window glass involves the fusing of a mixture of sand, SiO_2; calcium carbonate, $CaCO_3$; and sodium carbonate, Na_2CO_3. The process generates a gas. What is the identity of the gas?

(A) CO_2
(B) O_2
(C) Na_2O
(D) O_3

5. An aqueous mixture of H_2SO_4, $KMnO_4$, and $H_2C_2O_4$ reacts as follows:

$$3\ H_2SO_4(aq) + 2\ KMnO_4(aq) + 5\ H_2C_2O_4(aq) \rightarrow$$
$$K_2SO_4(aq) + 2\ MnSO_4(aq) + 10\ CO_2(g) + 8\ H_2O(l)$$

In one experiment, the initial mixture contained 4 moles of H_2SO_4, 2 moles of $KMnO_4$, and 5 moles of $H_2C_2O_4$. Approximately how many grams of sulfuric acid remain after the reaction?

(A) 100 g
(B) 200 g
(C) 300 g
(D) 150 g

6. Ag_2O decomposes when heated according to the following equation:

$$2\,Ag_2O \rightarrow 4\,Ag + O_2$$

How many moles of oxygen are formed by the decomposition of 3.0 moles of Ag_2O?

(A) 2.0 mol
(B) 1.0 mol
(C) 3.0 mol
(D) 1.5 mol

7. Light induces the following reaction in some photographic paper:

$$AgI(s) \rightarrow Ag(s) + I(s)$$

What type of process is this?

(A) Ionization reaction
(B) Oxidation reaction
(C) Lewis acid-base reaction
(D) Reduction reaction

8. How many grams of hydrogen are in 5 moles of sulfuric acid?

(A) 100 g
(B) 2.0 g
(C) 186 g
(D) 10 g

9. Determine the percent mass of each element in H_3PO_4.

(A) 38% hydrogen, 16% phosphorus, 46% oxygen
(B) 3.0% hydrogen, 32% phosphorus, 65% oxygen
(C) 4.0% hydrogen, 32% phosphorus, 64% oxygen
(D) 3.0% hydrogen, 47% phosphorus, 50% oxygen

10. What is the oxidation state of gold in cesium auride, $CsAu$?

(A) −1
(B) +1
(C) +3
(D) 0

11. Which of the following does not contain an element in a +1 oxidation state?

    (A) CuS
    (B) Cu_2S
    (C) $Ca(ClO)_2$
    (D) Hg_2Cl_2

12. The formula of pyruvic acid is $HC_3H_3O_3$. What is the mass percent of oxygen in pyruvic acid?

    (A) 30%
    (B) 55%
    (C) 41%
    (D) 60%

13. A buffer is 1.0 M in ammonia and 0.50 M in ammonium sulfate. What is the concentration of ammonium ions in the buffer?

    (A) 0.5 M
    (B) 1.5 M
    (C) 1.0 M
    (D) 2.0 M

14. How many milliliters of 0.40 M NaOH solution are necessary to neutralize 0.0500 L of 0.20 M H_2SO_4 solution?

    (A) 0.025 mL
    (B) 25.0 mL
    (C) 75.0 mL
    (D) 50.0 mL

15. The mass percent of oxygen in sucrose, $C_{12}H_{22}O_{11}$, is approximately:

    (A) 50%
    (B) 25%
    (C) 10%
    (D) 75%

16. The density of water is 1 g/mL. What is the molarity of water in pure water?

    (A) 55 M
    (B) 5.5 M
    (C) 18 M
    (D) 1 M

17. What is the oxidation state of carbon in formaldehyde (HCHO)?

(A) 0

(B) +4

(C) −4

(D) +2

18. What is the oxidation state of xenon in the perxenate ion (XeO_6^{4-})?

(A) 0

(B) −4

(C) −1

(D) +8

19. The following reaction illustrates the ability of hydrogen to serve as a reducing agent:

$$Re_2O_7(s) + 7\ H_2(g) \rightarrow 2\ Re(s) + 7\ H_2O(g)$$

A 1.0 kg sample of $Re_2O_7(s)$ is reacted with 1.0 kg of H_2. What is the maximum amount of Re formed?

(A) 0.77 kg

(B) 2.0 kg

(C) 0.50 kg

(D) 0.25 kg

20. The solution formed by mixing 100 mL of 3 M H_2SO_4 with 200 mL of 3 M KOH is identical to 300 mL of:

(A) 2 M K_2SO_4

(B) 3 M K_2SO_4

(C) 1 M K_2SO_4

(D) 1 M H_2SO_4 plus 2 M KOH

21. Dichlorine heptoxide, Cl_2O_7, is an unstable compound. Which of the following substances contains chlorine in the same oxidation state as the chlorine in dichlorine heptoxide?

(A) HCl

(B) $HClO_4$

(C) $HClO_3$

(D) Cl_2

22. What is the oxidation state of iron in Fe_3O_4?

    (A) +8/3
    (B) +3
    (C) +2
    (D) 0

23. Glucose, or blood sugar, is a carbohydrate. The molecular formula for glucose is $C_6H_{12}O_6$. Maltose, or malt sugar, is a disaccharide formed by linking two glucose molecules through a condensation reaction to form a glycoside linkage. What are the empirical formulas of glucose and maltose?

    (A) $C_6H_{12}O_6$ and $C_{12}H_{24}O_{12}$
    (B) $C_6H_{12}O_6$ and $C_{12}H_{22}O_{11}$
    (C) CH_2O and $C_{12}H_{22}O_{11}$
    (D) CH_2O and CH_2O

24. How many milliliters of 0.20 M $Ba(OH)_2$ solution are necessary to neutralize 0.0500 L of 0.20 M HCl solution?

    (A) 75 mL
    (B) 50 mL
    (C) 12 mL
    (D) 25 mL

25. What is the concentration of hydrogen ions in a solution made by mixing 250.0 mL of 0.10 M NaOH with 250.0 mL of 0.20 M HCl?

    (A) 0.050 M
    (B) 0.10 M
    (C) 0.010 M
    (D) 0.20 M

26. A sample contains 3.01×10^{23} of C_2H_5OH molecules. How much does the sample weigh?

    (A) 46 g
    (B) 23 g
    (C) 12 g
    (D) 92 g

27. What is the molality of a solution containing 16 g of CH_3OH in 500.0 g of water?

    (A) 2.0 m
    (B) 0.50 m
    (C) 1.0 m
    (D) 16 m

28. The oxidation states of the atoms in formaldehyde, CH_2O, are:

(A) $H = 0$, $C = +2$, and $O = -2$
(B) $H = +1$, $C = 0$, and $O = -2$
(C) $H = +1$, $C = -2$, and $O = +2$
(D) $H = -1$, $C = +4$, and $O = -2$

29. Approximately how many grams of phosphorus are in 1.5 moles of P_4?

(A) 31 g
(B) 130 g
(C) 180 g
(D) 15 g

30. What is the empirical formula of a compound containing 77.3% silver, 7.4% phosphorus, and 15.3% oxygen?

(A) Ag_3PO_4
(B) Ag_3PO_3
(C) $AgPO_3$
(D) $AgPO_4$

31. Arrange the following compounds in order of increasing oxidation number of bromine: $NaBrO_3$, $HBrO_2$, $Ca(BrO)_2$, and $HBrO_4$.

(A) $Ca(BrO)_2$, $HBrO_2$, $NaBrO_3$, and $HBrO_4$
(B) $HBrO_2$, $Ca(BrO)_2$, $NaBrO_3$, and $HBrO_4$
(C) $HBrO_4$, $NaBrO_3$, $HBrO_2$, and $Ca(BrO)_2$
(D) $Ca(BrO)_2$, $HBrO_2$, $HBrO_4$, and $NaBrO_3$

32. Which of the following is the percent composition of ethyl alcohol (C_2H_5OH)?

(A) 52% carbon, 11% hydrogen, 37% oxygen
(B) 52% carbon, 13% hydrogen, 35% oxygen
(C) 26% carbon, 13% hydrogen, 61% oxygen
(D) 19% carbon, 11% hydrogen, 70% oxygen

33. The formal charges of the atoms in the cyanide ion, CN^-, are:

(A) -1 for carbon and 0 for nitrogen
(B) 0 for carbon and -1 for nitrogen
(C) $-1/2$ for each atom
(D) $+2$ for carbon and -3 for nitrogen

34. Determine the molar concentration of a solution containing 0.20 moles of solute in 100.0 cm^3 of solution.

   (A) 20 M
   (B) 0.2 M
   (C) 2.0 M
   (D) 0.02 M

35. Determine the molality of a solution containing 5 moles of ethanol (C_2H_5OH) in 1,000 g of water.

   (A) 0.05 m
   (B) 0.005 m
   (C) 5000 m
   (D) 5 m

36. A solution was prepared by mixing 50.0 mL of 0.10 M Na_2CO_3(aq) with 50.0 mL of 0.10 M NaCl(aq). What is the concentration of Na^+(aq) in this solution?

   (A) 0.15 M
   (B) 0.10 M
   (C) 0.050 M
   (D) 0.075 M

37. A sample contains Avogadro's number of water molecules. How much does the sample weigh?

   (A) 6×10^{23} g
   (B) 18 g
   (C) 1 g
   (D) 17 g

38. In the following reaction, how many moles of H_2S form when 2.0 moles of HCl react with 0.5 moles of FeS?

$$2 \, HCl(g) + FeS(s) \rightarrow FeCl_2(s) + H_2S(g)$$

   (A) 2.5 mol
   (B) 2.0 mol
   (C) 0.5 mol
   (D) 1.0 mol

CHAPTER 2

Atomic Structure

Questions 39–44 refer to the following passage.

The rate of radioactive decay depends only upon the amount of radioisotope present in the sample. For this reason, the decay of radioisotopes is a perfect example of first-order kinetics.

Radioactive isotopes have found many uses in the medical profession. Today, there are two major uses for radioactive materials in medicine—diagnostic and therapeutic (treatment). In a diagnostic role, radioactive isotopes are useful as tracers to observe the structure and function of organs, while the therapeutic role involves treating a disease with ionizing radiation.

An example of a diagnostic function is the use of radioactive iodine-131 in the study of the function of the thyroid gland. Oral administration of I-131 leads to its concentration in the thyroid. It is possible to monitor this accumulation with a gamma-ray detector, and the information is displayed on a monitor. Another example is the use of thallium-201 to determine if an individual has heart diseases caused by narrowing of the arteries. The Tl-201 is injected into the bloodstream, where it binds strongly to heart muscle that is receiving a strong blood supply. Narrowed arteries do not allow sufficient blood flow to the heart, and the thallium does not bind. Monitors are used to detect the differences. Other tracer examples include iron-59, which is useful in measuring the rate of red blood cell formation, and phosphorus-32, which is used to locate tumors.

A type of imaging method, positron-emission tomography (PET), is used to observe brain structure and function. PET uses short-lived positron emitters such as N-13 ($t_{1/2}$ = 9.98 minutes), C-11 ($t_{1/2}$ = 20.4 minutes), F-18 ($t_{1/2}$ = 110 minutes), and O-15 ($t_{1/2}$ = 2.05 minutes). One of these isotopes is incorporated into a biological substance (such as glucose, water, or carbon dioxide), which is then injected into the patient. This substance then travels to the brain. There it emits positrons. Each positron reacts with an electron, giving off two gamma rays in opposite directions.

A gamma ray detector detects these pairs of gamma rays and over time creates an image of the tissue containing the positron emitter. PET can tell the difference between a normal brain and one of an Alzheimer's disease patient.

Other isotopes are used in the therapeutic (treatment) aspect of medicine. The most common disease treated in this fashion is cancer. Cancer cells are dividing rapidly and are more sensitive to radiation that interferes with the cell-division process than normal tissues are. This radiation is delivered to the cancerous site in a variety of ways. In the treatment of hyperthyroidism (overactive thyroid function), iodine-131 is injected into the body and accumulates in the thyroid gland. The radiation destroys the gland, while the short half-life, 8 days, means that the surrounding tissue undergoes minimal damage. Implanted beads of Au-198 or a mixture of Sr-90 and Y-90 has been useful in the treatment of breast and pituitary tumors. Other tumors may be treated by using highly focused beams of gamma radiation (from a cobalt-60 source) or x-rays. It is hoped that the rapidly dividing cancer cells may be quickly destroyed without damage to the surrounding healthy tissue.

39. The radioisotope iodine-131 is useful in studying the function of the thyroid gland. The isotope has a half-life of 8.0 days. What percent of the iodine-131 remains in a sample after 40 days?

    (A) 12.5%
    (B) 6.25%
    (C) 3.13%
    (D) 50.0%

40. A radioisotope undergoes decay with a half-life of 8 minutes. How long will it take for only 5% of the sample to remain?

    (A) 16 minutes
    (B) 35 minutes
    (C) 65 minutes
    (D) 8 minutes

41. The half-life of carbon-14 is about 5,700 years. Approximately how long will it take for 1 g of carbon-14 to decay to 0.23 g?

    (A) 5,700 years
    (B) 11,400 years
    (C) 17,100 years
    (D) 22,800 years

42. Which of the following radioactive nuclei is *least* likely to be harmful in contact with human skin?

    (A) ^{187m}Pb: half-life = 15.2 s, alpha emission; particle energy = 5.99 MeV
    (B) ^{230}Ac: half-life = 2.0 m, beta emission; particle energy = 1.4 MeV
    (C) ^{211}At: half-life = 7.21 h, electron capture; particle energy = 0 MeV
    (D) ^{232}Th: half-life = 1.4×10^{10} y, alpha emission; particle energy = 2.83 MeV

43. A laboratory accident resulted in the release of radioactive radon-222 ($t_{1/2}$ = 3.8 days). The initial radiation level was 1×10^{-5} μCi/mL (microcuries per milliliter). Even with protective respiratory gear, it is not safe to enter the laboratory until the radiation level drops below 1×10^{-6} μCi/mL. What is the least number of days required for the radiation level to drop below 1×10^{-6} μCi/mL?

   (A) 3.8 days
   (B) 10 days
   (C) 12 days
   (D) 16 days

44. The half-life of hydrogen-3 is about 14 years. What is the approximate amount remaining after 8 years?

   (A) 67%
   (B) 50%
   (C) 90%
   (D) 25%

Questions 45–53 refer to the following passage.

Over 270 atomic nuclei are stable. An even larger number of nuclei are unstable. An unstable nucleus will eventually transform (decay) into another more-stable nucleus. **Radioactive decay** is the process by which an unstable nucleus transforms into another nucleus. There are a number of radioactive decay processes. However, only a few are common.

The most common types of radioactive decay are alpha emission, beta emission, and gamma emission. **Alpha emission** is the radioactive decay mode in which a nucleus emits an alpha particle. **Beta emission** is the radioactive decay mode in which a nucleus emits a beta particle. **Gamma emission** is the radioactive decay mode in which a nucleus emits a gamma ray. In all three of these cases, something comes out of the nucleus of an atom.

Three additional types of radioactive decay are important, even though they are less common. These are electron capture, positron decay, and spontaneous fission. **Electron capture** is the radioactive decay mode in which a nucleus absorbs (captures) an electron. **Positron emission** is the radioactive decay mode in which a nucleus emits a positron particle. **Spontaneous fission** is the radioactive decay mode in which a nucleus splits into two or more different nuclei. **Neutron emission** is the radioactive decay mode in which a nucleus emits a neutron ($_0^1 n$).

Electron capture, unlike other decay modes, involves something entering the nucleus. Positron emission, like the three more common decay modes, involves the emission of a particle from the nucleus. Spontaneous fission, which has no specific symbol, does not involve anything entering the nucleus.

The nucleus formed by a radioactive decay process may be in an excited state, a high energy state. Excited-state nuclei, also known as metastable nuclei, typically emit one or more gamma rays in order to reach the ground state. A well-known example of such a nucleus is technetium-99m or ^{99m}Tc (the *m* stands for metastable).

45. The bombardment of thorium-232 with alpha particles results in the absorption of the particle and the emission of a free neutron. What nuclide forms?

    (A) ^{235}Th
    (B) ^{236}U
    (C) ^{235}U
    (D) ^{236}Th

46. What radioactive decay mode is involved in the conversion of ^{118}Xe to ^{118}I?

    (A) β^- decay
    (B) β^+ decay
    (C) α decay
    (D) Gamma emission

47. The isotope ^{100}Pd undergoes electron capture. Which isotope does this process produce?

    (A) ^{100}Ag
    (B) ^{100}Rh
    (C) ^{96}Ru
    (D) ^{99}Pd

48. An isotope of an element undergoes a series of radioactive decay processes to produce a different isotope of the same element. If the first step was an alpha decay, how many beta decay steps were there?

    (A) Four
    (B) One
    (C) Zero
    (D) Two

49. Carbon-14 is a beta emitter. What isotope does the beta emission of this isotope produce?

    (A) ^{14}N
    (B) ^{12}C
    (C) ^{14}Si
    (D) ^{13}C

50. If thorium-230 undergoes two beta decays and an alpha decay, what isotope forms?

    (A) Thorium-226
    (B) Radium-226
    (C) Thorium-232
    (D) Actinium-226

51. Analysis of the air in a container with an unknown radioactive isotope shows a high concentration of helium gas. What type of radioactive decay does the unknown radioisotope experience?

    (A) Positron emission
    (B) Beta emission
    (C) Gamma decay
    (D) Alpha emission

52. If a radioactive atom emits two alpha particles, by how many units does its mass change?

    (A) 8
    (B) 4
    (C) 2
    (D) 0

53. The radioactive decay series of ^{232}Th ends with ^{208}Pb. How many α and β particles are emitted during this decay?

    (A) 6 α and 7 β particles
    (B) 7 α and 7 β particles
    (C) 8 α and 6 β particles
    (D) 7 α and 6 β particles

54. Which of the following gaseous atoms has two unpaired electrons in the ground state?

    (A) Calcium
    (B) Vanadium
    (C) Nickel
    (D) Aluminum

55. The actual mass of an isotope is less than the sum of the masses of the constituents by an amount known as the:

    (A) Atomic weight
    (B) Atomic mass
    (C) Mass defect
    (D) Molar mass

56. Which element might have the excited-state electron configuration $1s^2 2s^2 2p^5 3s^2 3d^1$?

   (A) Al
   (B) Mg
   (C) Na
   (D) Ar

57. Which is the only noble gas that does *not* have an ns^2np^6 outer shell electron configuration?

   (A) He
   (B) Ne
   (C) Ar
   (D) Rn

58. Which element has the following electron configuration: $[Kr]5s^2 4d^7$?

   (A) Ir
   (B) Co
   (C) Rh
   (D) Tc

59. Most periodic tables include certain information about an element. In addition to the location of the element on the periodic table, the entry for chlorine contains the letter Cl, the number 17, and the number 35.453 in the "box" for the element. What is the mass number of chlorine?

   (A) 17
   (B) 35.453
   (C) 52.453
   (D) There is insufficient information.

60. Assuming equal energy of the radiation, which radioactive particle shows the greatest deviation from its path when entering an electrical field?

   (A) α particle
   (B) β^- particle
   (C) Proton
   (D) γ ray

61. Which of the following compounds has oxygen in the highest oxidation state?

   (A) OF_2
   (B) H_2O
   (C) H_2O_2
   (D) KO_2

62. Which of the following is isoelectronic with the iron ion in Fe_2O_3?

(A) Cr^{2+}
(B) Co^{2+}
(C) Mn^{2+}
(D) Fe^{2+}

63. The n = 4 shell can hold a maximum of:

(A) 2 electrons
(B) 8 electrons
(C) 18 electrons
(D) 32 electrons

64. Bohrium is a man-made element that fits on the periodic table in the manganese family (below Re). One of the electrons on a bohrium atom has a quantum number of –3. Which subshell holds this electron?

(A) 7s
(B) 6d
(C) 5f
(D) 7p

65. The p orbitals on a nitrogen atom are degenerate. What does *degenerate* mean?

(A) The energies of the orbitals are the same.
(B) There are three orbitals in the p orbital set.
(C) The orbitals can form pi (π) bonds.
(D) The orbitals can hybridize.

66. Iodine-131 is a beta emitter. The nuclear accident at Fukushima, Japan, released quantities of radioactive iodine-131 into the atmosphere. The product of the radioactive decay of iodine-131 is:

(A) ^{131}Xe
(B) ^{130}I
(C) ^{127}Sb
(D) ^{131}Te

67. Which of the following qualifies as a Bohr atom?

(A) He^+
(B) H^-
(C) H^+
(D) Li^+

68. Certain numbers of protons or neutrons impart additional stability to a nucleus. These numbers are the so-called "magic numbers" (2, 8, 20, 28, 50, 82, 126 . . .). Nuclei with a magic number of either protons or neutrons are less likely to be radioactive (doubly magic are even more stable). Which of the following nuclei is least likely to be radioactive?

(A) ^{8}Li
(B) ^{12}B
(C) ^{8}Be
(D) ^{15}N

69. Certain numbers of protons or neutrons impart additional stability to a nucleus. These numbers are the so-called "magic numbers" (2, 8, 20, 28, 50, 82, 126 . . .). Nuclei with a magic number of either protons or neutrons are less likely to be radioactive (doubly magic are even more stable). Which of the following is a doubly magic nucleus?

(A) ^{57}Fe
(B) ^{118}Sn
(C) ^{208}Pb
(D) ^{17}O

70. Stable light isotopes normally have approximately equal numbers of protons and neutrons. Stable heavier isotopes normally have more neutrons than protons. Isodiapheres are isotopes of different elements with the same number of excess neutrons over protons. Which of the following pairs represent isodiapheres?

(A) ^{238}U and ^{234}Th
(B) ^{208}Pb and ^{202}Hg
(C) ^{128}Xe and ^{130}Ba
(D) ^{16}O and ^{22}Ne

71. Strontium-90 is a radioactive isotope produced in nuclear explosions. It has a half-life of 29 years and emits beta particles. The *best* explanation of why strontium-90 is a beta emitter is that the:

(A) Neutron-electron ratio is too high
(B) Neutron-proton ratio is too low
(C) Neutron-proton ratio is not 1
(D) Neutron-proton ratio is too high

72. The flame test for lithium is a brilliant crimson in color. Which of the following is the best explanation?

(A) Energy is emitted, and there is a bright band in the atomic spectrum.
(B) Energy is emitted, and there is a dark band in the atomic spectrum.
(C) Energy is absorbed, and there is a bright band in the atomic spectrum.
(D) Energy is absorbed, and there is a dark band in the atomic spectrum.

73. Which of the following species has electrons in the highest energy subshell? Assume all species are in their ground states.

(A) K
(B) Ca^{2+}
(C) Cl^-
(D) Ar

74. Cobalt may form a Co^{3+} ion. What is the ground-state electron configuration of Co^{3+}?

(A) $[Ar]4s^23d^7$
(B) $[Ar]4s^23d^4$
(C) $[Ar]$
(D) $[Ar]3d^6$

75. Based upon the electron configurations listed, which of the following ions experiences the least attraction to a magnet?

(A) Ca^+ $[Ar]4s^1$
(B) Fe^{3+} $[Ar]3d^5$
(C) Cu^+ $[Ar]3d^{10}$
(D) Ga^{2+} $[Ar]3d^{10}4s^1$

76. Which is a representation of the electron configuration of silicon in an excited state?

(A) $1s^22s^22p^63s^23p^14s^1$
(B) $1s^22s^22p^63s^23p^24s^1$
(C) $1s^22s^22p^63s^23p^14s^2$
(D) $1s^22s^22p^63s^23p^3$

77. A nuclear fusion process converted 2.0×10^{-28} g to energy. How much energy was released?

(A) 1.8×10^{-11} J
(B) 1.8×10^{-14} J
(C) 6.0×10^{-20} J
(D) 6.0×10^{23} J

78. If two nuclei fuse to produce another isotope and release energy, what can be said about the exact mass of the nucleus formed?

    (A) It is equal to the sum of the two nuclei.
    (B) It is greater than the sum of the two nuclei.
    (C) It is less than the sum of the two nuclei.
    (D) It is equal to the average of the two nuclei.

79. A valence electron on a sodium atom might have which of the following sets of quantum numbers?

    (A) 3, 1, 0, −½
    (B) 3, 0, 1, +½
    (C) 2, 0, 0, +½
    (D) 3, 0, 0, +½

80. An electron on an atom has the following quantum numbers: 5, 2, 0, +½. How many other electrons on the same atom can have this set of quantum numbers?

    (A) 3
    (B) 2
    (C) 1
    (D) 0

81. How many sp² hybridized carbon atoms are in the following molecule?

(A) 9
(B) 5
(C) 15
(D) 3

82. Three of the following compounds are colorless. Which is the exception?

(A) ZnO
(B) TiO_2
(C) Fe_2O_3
(D) Sc_2O_3

83. A nuclear accident released quantities of radioactive iodine-131 into the atmosphere. Iodine is an essential element for humans. The concern at the time was that humans would absorb the radioactive iodine along with natural (nonradioactive) iodine. One treatment to reduce the amount of iodine-131 absorbed would be to:

(A) Have potential victims ingest diuretics to maximize loss of iodine-131 contaminated water
(B) Have potential victims boil possibly contaminated food and water to increase the radioactive decay of iodine-131
(C) Have potential victims ingest silver nitrate tablets to precipitate the iodine as silver iodide
(D) Have potential victims ingest large amounts of nonradioactive iodine

CHAPTER 3

Bonding and the Periodic Table

Questions 84–88 refer to the passage below.

When studying the bonding behavior of the representative elements, a diagrammatic method simplifies keeping track of the behavior of the valence electrons. This method takes advantage of Lewis structures built from a combination of Lewis symbols. A **Lewis symbol** is the chemical symbol of an element with some indication of the valence electrons. We may use dots, circles, or any other symbol placed around a central chemical symbol of an element to identify the valence electrons. We can write a Lewis structure for every atom, ion, or compound. The Lewis structure immediately gives us information about the valence electrons and the bonding. It also gives us information that allows us to predict the stoichiometry of compounds.

84. Which of the following is the correct Lewis structure for O_2?

    (A) $\ddot{O} = \ddot{O}$

    (B) $\ddot{O} - \ddot{O}$

    (C) $:O = O:$

    (D) $:\ddot{O} - \ddot{O}:$

85. Oxygen molecules will bind to the iron in hemoglobin. When this occurs, the oxygen molecule is acting as:

    (A) An oxidizing agent

    (B) A Lewis acid

    (C) A Lewis base

    (D) A reducing agent

86. The *best* way of describing the binding of oxygen to the iron in hemoglobin is the:

    (A) Formation of a dipole-dipole interaction
    (B) Formation of an ionic bond
    (C) Formation of a metallic bond
    (D) Formation of a coordinate covalent bond

87. When an oxygen molecule binds to the iron in hemoglobin, the iron becomes six-coordinate. The approximate geometry of the atoms around the iron is:

    (A) Trigonal bipyramidal
    (B) Square planar
    (C) Tetrahedral
    (D) Octahedral

88. Which of the following amino acid residues is capable of forming a coordinate covalent bond to an iron ion?

(A) Aspartic acid

$$H_2N-CH-C-OH$$

with structure:

$$\begin{array}{c} O \\ \| \\ H_2N-CH-C-OH \\ | \\ CH_2 \\ | \\ C=O \\ | \\ OH \end{array}$$

(B) Valine

$$\begin{array}{c} O \\ \| \\ H_2N-CH-C-OH \\ | \\ CH-CH_3 \\ | \\ CH_3 \end{array}$$

(C) Alanine

$$\begin{array}{c} O \\ \| \\ H_2N-CH-C-OH \\ | \\ CH_3 \end{array}$$

(D) Glycine

$$\begin{array}{c} O \\ \| \\ H_2N-CH-C-OH \\ | \\ H \end{array}$$

89. How does the atomic radius of potassium relate to the atomic radius of arsenic?

(A) Potassium is smaller because the effective nuclear charge is smaller.
(B) Potassium is larger because the effective nuclear charge is larger.
(C) Potassium is larger because the effective nuclear charge is smaller.
(D) Potassium is smaller because the effective nuclear charge is larger.

90. In the Lewis structure of $MgCl_2$, how many electrons are shared?

    (A) 0
    (B) 1
    (C) 2
    (D) 8

91. How does the third ionization energy of Ca compare to the third ionization energy of Sc?

    (A) The value for Ca is much higher.
    (B) The value for Ca is much lower.
    (C) The values are the same.
    (D) Unrelated because Sc is a transition element and Ca is not.

92. An excited-state potassium atom emits light of characteristic wavelengths. These emitted wavelengths are due to:

    (A) Excited-state electrons moving to higher energy levels
    (B) Excited-state electrons dropping to lower energy levels
    (C) Transitions within the potassium nucleus
    (D) Excited-state potassium atoms interacting with ground-state potassium atoms

93. It is possible to solidify helium at 1 K under a pressure of 26 atm. What is the specific reason why this occurs?

    (A) There are temporary dipole-dipole attractions between the helium atoms.
    (B) At this temperature, gravitational forces are sufficient to hold the atoms together.
    (C) At this temperature, it is possible to form weak covalent bonds between the helium atoms.
    (D) The helium atoms convert to a collection of alpha particles and free electrons, which are held together by electrostatic forces.

94. Which of the following represents the predicted relationship between the radii of the species?

    (A) $La^{3+} < Ba^{2+} < Xe < Te^{2-}$
    (B) $Xe < Te^{2-} < Ba^{2+} < La^{3+}$
    (C) $Te^{2-} < Xe < Ba^{2+} < La^{3+}$
    (D) $La^{3+} > Ba^{2+} > Xe > Te^{2-}$

95. The rate of the passive diffusion of ions across a cell membrane is inversely proportional to the radius of the ion. Which of the following ions has the slowest rate of passive diffusion?

 (A) F^-
 (B) K^+
 (C) Ba^{2+}
 (D) Mg^{2+}

96. Many properties of the elements show a general periodic trend. What is the general trend in the metallic character of the elements?

 (A) Decreases to the left and downward
 (B) Increases to the left and downward
 (C) Decreases to the left and increases downward
 (D) Increases to the left and decreases downward

97. Why is the second electron affinity of an atom significantly greater than the first electron affinity?

 (A) This makes an electron capture process more likely.
 (B) This completes the stable electron configuration.
 (C) The second electron is held more tightly than the first.
 (D) There is repulsion of like charges.

98. Why is the ionization energy of sodium lower than that of chlorine?

 (A) The valence electron of sodium is closer to the nucleus and experiences a lower effective nuclear charge.
 (B) The valence electron of sodium is farther from the nucleus and experiences a lower effective nuclear charge.
 (C) The valence electron of sodium is farther from the nucleus and experiences a higher effective nuclear charge.
 (D) The valence electron of sodium is closer to the nucleus and experiences a higher effective nuclear charge.

99. The following is a Lewis structure for nitrous acid. What is the correct sequence (left to right) of the formal charges?

$$H-\overset{..}{\underset{..}{O}}-\overset{..}{N}=\overset{..}{O}:$$

 (A) +1, −2, +3, −2
 (B) 0, 0, 0, 0
 (C) +1, −1, +2, −2
 (D) −1, −2, +5, −2

100. Referring to the Lewis structure for nitrous acid in the previous question, what is the correct hybridization (left to right) of the oxygen atoms?

(A) sp^3 and sp
(B) sp and sp^2
(C) sp^3 and sp^3
(D) sp^3 and sp^2

101. Referring to the Lewis structure for nitrous acid in question 99, how do the lengths of the nitrogen-oxygen bonds compare when nitrous acid loses a hydrogen ion to form a nitrite ion?

(A) They are equal.
(B) The bond to the oxygen on the left is longer.
(C) The bond to the oxygen on the left is shorter.
(D) The relative lengths are always changing.

102. The following is the structure of the pyruvate ion:

The approximate geometry around the carbon atoms bonded to oxygen (left to right) is:

(A) Trigonal pyramidal and trigonal planar
(B) Trigonal planar and trigonal pyramidal
(C) Trigonal planar and trigonal planar
(D) Tetrahedral and tetrahedral

103. What is the shape of the carbonate ion?

(A) Tetrahedral
(B) Trigonal planar
(C) Linear
(D) Octahedral

104. Hydrogen chloride gas will react with ammonia gas to produce ammonium chloride. At room temperature, ammonium chloride will exist as what phase?

(A) Gas
(B) Plasma
(C) Liquid
(D) Solid

105. Electrolytes are compounds that produce ions when dissolving in a solvent. Which of the following compounds are *not* electrolytes when dissolved in water?

I.   Ethanol
II. Hydrogen bromide
III. Carbon monoxide

(A) I and III only
(B) I and II only
(C) II only
(D) I only

106. Which of the following is a free radical?

(A) NO_2^-
(B) NH_3
(C) NO_3^-
(D) NO

107. All the H–C–H angles in the methyl radical, CH_3, are 120°. Where is the unpaired electron?

(A) In an sp^3 hybridized orbital
(B) In an sp^2 hybridized orbital
(C) In an unhybridized p-orbital
(D) Unknown

108. Which of the following does NOT have a trigonal planar orbital geometry?

(A) NO_3^-
(B) SO_3
(C) NH_3
(D) BF_3

109. Glass is primarily silicon dioxide. Hydrofluoric acid is the only common acid that will react with glass. The reaction produces water and silicon tetrafluoride. Which of the following statements about silicon tetrafluoride is true?

    (A) It is polar.
    (B) It is nonpolar.
    (C) It is ionic.
    (D) It is explosive.

110. The nitrogen atom in nitric acid (HNO_3) is bonded to all three oxygen atoms. This means that the:

    (A) Nitrogen forms one double bond and two single bonds to the oxygen atoms
    (B) Nitrogen forms three single bonds to the oxygen atoms
    (C) Nitrogen forms two double and one single bond to the oxygen atoms
    (D) Nitrogen atom has a nonbonded pair of electrons

111. Chelation therapy is useful in the treatment of heavy-metal poisoning. The process involves treatment with chelating agents such as BAL (British anti-Lewisite). Chelating agents are ligands that bind to heavy metal ions through the donation of electron pairs to the metal ion (cation). The bond to the metal is a very polar coordinate covalent bond. Which of the following statements is true?

    (A) Calcium ions would be good ligands.
    (B) Binding to the chelating agent increases the positive charge on the metal.
    (C) Ethane should be a good ligand.
    (D) The ligand is a Lewis base.

112. What is the strongest intramolecular force present in the bicarbonate ion, $HCO_3{}^-$?

    (A) Ionic bonding
    (B) Covalent bonding
    (C) Hydrogen bonding
    (D) Dipole-dipole forces

113. Hydrogen bonds are relatively strong intermolecular forces that are important for many of the properties of water, the double helix of DNA, and the secondary structure of proteins. The hydrogen bonds in water are stronger than the intermolecular interactions of all of the following *except*:

(A) The intermolecular forces in NaCl
(B) The intermolecular forces in ammonia
(C) The intermolecular forces in CH_3F
(D) The intermolecular forces in CH_3CH_3

114. The O–N–O bond angle in the nitrite ion is:

(A) Indeterminate
(B) Ideal
(C) Greater than ideal
(D) Less than ideal

115. What is the shape of ClF_4^-?

(A) Square planar
(B) Tetrahedral
(C) Square pyramid
(D) Irregular tetrahedron

116. A burette is a piece of laboratory glassware designed to measure precisely the volume of solution delivered during a titration. Due to capillary action, the solution in the burette forms a meniscus (curved surface) as the solution near the edges climbs up the side of the tube. The cause of the capillary action is the strong intermolecular forces between the glass walls and the solution in the tube. In order to measure the quantity of liquid dispensed, it is necessary to take readings before and after the delivery of a quantity of solution and take the difference. To get accurate readings:

(A) Both need to be at the top of the meniscus
(B) Both need to be at the bottom of the meniscus
(C) Both need to be at the average between the top and the bottom of the meniscus
(D) It does not matter if the top or the bottom is used, as long as the choice is consistent

117. The cyanate ion, OCN^-, and the fulminate ion, CNO^-, are isomers. In each case, the arrangement of the atoms is as shown in the formula. Why is the fulminate ion less stable?

    (A) There is less resonance stabilization of the fulminate ion because there are no multiple bonds.
    (B) There are more π-bonds in the cyanate ion, making it more stable.
    (C) There is less resonance stabilization and an unfavorable distribution of formal charges.
    (D) The fulminate ion is a very reactive free radical, and the cyanate ion has all electrons paired.

118. Electronic transitions in water molecules require photons with energies in the ultraviolet region of the spectrum. Vibrational transitions in water molecules require less energy. Photons causing vibrational transitions have energies in which region of the spectrum?

    (A) Ultraviolet
    (B) X-ray
    (C) Gamma-ray
    (D) Infrared

119. Infrared spectroscopy examines the vibrational spectra of molecules. It is possible to use infrared spectroscopy to examine solids, liquids, and gases. Which of the following gases has *no* vibrational spectrum?

    (A) Xe
    (B) H_2O
    (C) HCN
    (D) NO

120. Which of the following molecules is the least polar?

    (A) HCl
    (B) NH_3
    (C) SeF_4
    (D) CF_4

121. Which of the following cannot participate in intermolecular hydrogen bonding?

    (A) CH_3OH
    (B) CHF_3
    (C) H_2SO_4
    (D) HN_3

122. Phosphorus forms a number of compounds with the halogens. One phosphorus halide is PF_5. Which of the following statements is true concerning PF_5?

(A) The formal charge on phosphorus is 0, and the molecular geometry is trigonal bipyramidal.

(B) The formal charge on phosphorus is +5, and the molecular geometry is trigonal bipyramidal.

(C) The formal charge on phosphorus is 0, and the molecular geometry is square pyramidal.

(D) The formal charge on phosphorus is +5, and the molecular geometry is square pyramidal.

123. Which of the following is expected to have the lowest second ionization energy? Assume all species are in their ground states.

(A) Ba

(B) Cs

(C) Ca

(D) H

124. Which of the following properties is unlikely to be true for a metal such as copper?

(A) High ductility

(B) High malleability

(C) High electrical conductivity

(D) High brittleness

125. Boron halides are capable of reacting with halide ions via the following general reaction:

$$BX_3 + X^- \rightarrow BX_4^-$$

During this reaction, the boron halide behaves as a:

(A) Lewis acid, and the geometry changes from trigonal planar to tetrahedral

(B) Lewis acid, and the geometry changes from trigonal pyramidal to tetrahedral

(C) Lewis base, and the geometry changes from trigonal planar to tetrahedral

(D) Lewis base, and the geometry changes from trigonal pyramidal to tetrahedral

126. Which of the following series indicates a correct order of increasing polarity?

    (A) F–F < N–F < C–F

    (B) F–F < C–F < N–F

    (C) N–F < C–F < F–F

    (D) C–F < N–F < F–F

127. The mineral sphalerite, ZnS, has a structure with each zinc atom tetrahedrally surrounded by sulfur, and each sulfur atom tetrahedrally surrounded by zinc. Replacing both the zinc and sulfur with carbon will result in the diamond structure. This means that:

    (A) London dispersion forces replace ionic bonds

    (B) Ionic bonds replace covalent bonds

    (C) Covalent bonds replace ionic bonds

    (D) Dipole-dipole forces replace covalent bonds

128. Which of the following is the correct order of increasing melting point?

    (A) $HCl < Cl_2 < H_2O < NaCl$

    (B) $Cl_2 < HCl < H_2O < NaCl$

    (C) $HCl < NaCl < Cl_2 < H_2O$

    (D) $Cl_2 < NaCl < HCl < H_2O$

129. Antifreeze (ethylene glycol) is effective at lowering the freezing point of water. The addition of 2 moles of ethylene glycol to 1 L of water will lower the freezing point of the solution to $-3.8°C$. Ideally, what will be the freezing point of a solution made by adding 73 g of HCl to 1 L of water?

    (A) $-7.6°C$

    (B) $-3.8°C$

    (C) $0.0°C$

    (D) $-11.4°C$

130. Which of the following has the lowest ionization energy?

    (A) Calcium

    (B) Lithium

    (C) Potassium

    (D) Magnesium

131. What general trend is true for any period on the periodic table?

(A) Electronegativity decreases and electron affinity increases with increasing atomic number.

(B) Electronegativity and electron affinity decrease with increasing atomic number.

(C) Electronegativity and electron affinity increase with increasing atomic number.

(D) Electronegativity increases and electron affinity decreases with increasing atomic number.

132. All of the following geometries for compounds can be nonpolar *except*:

(A) Bent

(B) Linear

(C) Tetrahedral

(D) Octahedral

133. Why is it easier for sodium to form cations than anions?

(A) Sodium has a high electron affinity.

(B) Sodium has a high ionization energy.

(C) Sodium has a low ionization energy.

(D) Sodium can complete its outer shell by gaining an electron.

134. Potassium metal reacts explosively with all of the following *except*:

(A) Fluorine gas

(B) Argon gas

(C) Water vapor

(D) Oxygen gas

135. Channels are present in cell membranes to allow the selective transport of ions through the membrane. Size is one of the factors that allow the selective transport of the ions. Membranes are less permeable to larger ions than to smaller ions. Which of the following statements is correct?

(A) Sodium ions and chloride ions transport at the same rate because they are the same size.

(B) Smaller sodium ions transport more readily than larger potassium ions.

(C) Calcium ions transport more slowly because they are larger than potassium ions.

(D) Smaller potassium ions transport more readily than larger sodium ions.

136. Which of the following elements has the lowest electron affinity?

   (A) Sulfur
   (B) Chlorine
   (C) Neon
   (D) Fluorine

137. The electron configuration of phosphorus is $[Ne]3s^2 3p^3$. What is the electron configuration of a phosphide ion?

   (A) $[Ne]3s^2 3p^1$
   (B) $[Ne]3s^2 3p^4$
   (C) $[Ar]$
   (D) $[Ne]$

138. In which of the following resonance forms of the cyanate ion does the nitrogen atom have a formal charge of 0?

   (A) $\left[:\ddot{O} - C \equiv N: \right]^-$

   (B) $\left[:O \equiv C - \ddot{N}: \right]^-$

   (C) $\left[\ddot{O} = C = \ddot{N} \right]^-$

   (D) $\left[\ddot{O} - C - \ddot{N} \right]^-$

139. The following is the structure of the pyruvate ion:

   Which of the following statements about the pyruvate ion is true?

   (A) Upon lowering the pH, the product will be less soluble.
   (B) Upon lowering the pH, the product will be more soluble.
   (C) The longest carbon-oxygen bond is to the oxygen atom in the upper right of the structure.
   (D) The carbon-oxygen bonds on the left side of the structure are of unequal lengths.

140. Which of the following molecules adopts a square pyramidal geometry?

(A) Sulfur hexafluoride
(B) Antimony pentafluoride
(C) Xenon tetrafluoride
(D) Iodine pentafluoride

141. Sulfur trioxide will dissolve in water to produce sulfuric acid. During this process, the geometry around the sulfur atom changes from:

(A) Trigonal planar to tetrahedral
(B) Trigonal pyramidal to tetrahedral
(C) Trigonal pyramidal to trigonal pyramidal
(D) Trigonal planar to trigonal planar

142. Mercury and mercury compounds are toxic. In general, nonpolar dimethyl mercury is more toxic than ionic compounds such as mercury sulfide. Why is dimethyl mercury more toxic?

(A) Nonpolar dimethyl mercury can cross the hydrophobic cell membrane more easily than ionic mercury.
(B) Mercury ions can cross the hydrophobic cell membrane more easily than dimethyl mercury.
(C) Most ionic mercury compounds are soluble and are easily eliminated.
(D) The high pH in the small intestine enhances the absorption of dimethyl mercury.

143. Which of the following compounds will have the lowest melting point?

(A) CH_3CH_2COOH
(B) $CH_3CH_2CH_2OH$
(C) CH_3CH_2CHO
(D) $CH_3CH_2CH_2NH_2$

CHAPTER 4

Phases

Questions 144–153 refer to the following passage.

In many ways, water has very unusual properties. For this reason, it is not always best to think of water as a typical liquid. The highly polar water molecules are capable of hydrogen bonding. Most liquids do not form hydrogen bonds. Liquids such as liquid ammonia do not form as many hydrogen bonds as water. An ammonia molecule, with one lone pair, can attract one hydrogen atom from an adjacent ammonia molecule to form a hydrogen bond. Water, on the other hand, has two lone pairs, which can attract one hydrogen atom from each of two separate water molecules to form two hydrogen bonds.

The specific heat of water is 4.18 J/g°C, an unusually high value. It requires a large amount of heat energy to warm water. In addition, it is necessary to remove a significant amount of heat to cool water. For this reason, the temperature of the water in the oceans does not change easily. The volume of water in the oceans is sufficient to moderate the temperature of the Earth. The moon receives the same amount of energy from the sun, and so one would expect the temperatures to be about the same as on the Earth. The temperature of the moon varies from near –150°C to above the boiling point of water. This temperature range is unlike that on the Earth because the moon has very little available liquid or gaseous water to moderate the temperature change. To a lesser extent, water vapor in the atmosphere serves a similar purpose.

The density of liquid water is approximately 1.0 g/cm^3. The density is slightly lower than this value at temperatures above or below 4°C. The density of most liquids decreases as the temperature increases because the increased kinetic energy of the molecules pushes the particles away from each other. This is true for water above 4°C. Water is one of the few substances where the density of the solid form is lower than that of the liquid form. Bismuth metal is another example.

The strong hydrogen bonding means that the heat of vaporization of water is very high. It takes approximately 40 kilojoules to vaporize a mole of water. Even with this high value, it is necessary to evaporate massive amounts of water to supply the water vapor necessary for precipitation. The molecules of water vapor contain a large amount of heat energy. The evaporation of water also helps to moderate the Earth's temperature. The higher amount of energy is apparent when comparing the energy of liquid water at the normal boiling point to the amount of energy contained in water vapor at the same temperature. If you spill one teaspoon of boiling water on your hand, about 1,000 joules of heat will be available to burn your hand. The same mass of steam would supply over 10,000 joules of heat.

The strong hydrogen bonding in water leads to a high viscosity and surface tension. The high viscosity is not easy to observe, but is apparent if you try walking through the water in the shallow end of a swimming pool. Everyone has seen drops of water spilled on a surface. These drops tend to draw into rounded beads. This is a consequence of the high surface tension of water. Other liquids, such as gasoline, do not draw into well-rounded beads.

144. The heat of vaporization of water is unusually high for a small molecule. The main reason the value is high is because:

   (A) There are strong hydrogen bonds between the molecules.
   (B) The O–H bonds in water are very strong.
   (C) The water is a very light molecule.
   (D) The ionic bonds between H^+ and OH^- are strong.

145. The heat of fusion for ice is about 330 J/g, and the heat of vaporization of water is 2,260 J/g. What is the heat of deposition of water?

   (A) Impossible to determine
   (B) +2,590 J/g
   (C) +1,930 J/g
   (D) –2,590 J/g

146. The heat of fusion for ice is 6.0 kJ/mol, and the molar heat capacity of water is 75 J/mol°C. What is the temperature change when 16.0 kJ of heat is added to 54 g of ice?

   (A) 80°C
   (B) 0°C
   (C) 40°C
   (D) 60°C

147. A simplified phase diagram for water is:

The pressure and temperature at the normal boiling point are:

(A) P_3 and T_3
(B) P_1 and T_1
(C) P_4 and T_4
(D) P_2 and T_2

148. Below is the phase diagram for water:

A sample of water vapor at T_1 is compressed at constant temperature. The sequence of phase changes is:

(A) Solid → liquid → gas
(B) Gas → liquid → solid
(C) Gas → solid → liquid
(D) No phase change occurs

149. Water boils at a lower temperature at the top of a mountain than at sea level. The best explanation for this fact is that:

    (A) Ultraviolet light breaks the H–O bonds.
    (B) There is more solar energy available at higher altitudes.
    (C) The vapor pressure of water decreases with increasing altitude.
    (D) The atmospheric pressure is lower.

150. Which of the following is the chief factor leading to the density of ice being less than the density of water?

    (A) Hydrogen bonding
    (B) Formation of covalent bonds between the water molecules
    (C) The energy change when water freezes
    (D) The lower entropy of the solid

151. When water freezes, there is a nearly 10% increase in volume. This increase in volume may result in the bursting of the cooling system in an automobile or the cracking of a sidewalk in the winter. Which of the following substances would be the *least* effective at preventing the cracking of sidewalks? Assume one mole of each is available.

    (A) $C_6H_{12}O_6$
    (B) NaCl
    (C) $(NH_4)_2SO_4$
    (D) Na_3PO_4

152. The following is a simplified phase diagram of water.

What process occurs when heating a sample of water from −10°C to +10°C at a pressure of 4×10^{-3} atm?

(A) Deposition
(B) Sublimation
(C) Melting
(D) Boiling

153. The specific heat of water is 4.2 J/g°C. What is the heat capacity of 200 g of water?

(A) 840 J/g°C
(B) 840 J/°C
(C) 420 J/g°C
(D) 420 J/°C

154. Which best describes the changes occurring during spontaneous deposition?

(A) Heat is absorbed and the entropy increases.
(B) Heat is released and the entropy decreases.
(C) Heat is released and the entropy increases.
(D) Heat is absorbed and the entropy decreases.

155. Which of the following is normally the largest for a nonpolar molecule?

    (A) Heat of vaporization
    (B) Heat of sublimation
    (C) Heat of fusion
    (D) For a nonpolar substance, all are the same.

156. A substance is a solid at STP. Increasing the temperature at constant pressure may eventually cause:

    (A) Solidification
    (B) Deposition
    (C) Sublimation
    (D) Condensation

157. The normal boiling point of nitrogen is 77 K. However, it is possible to have liquid nitrogen as a stable phase at temperatures above 77 K. Why is this possible?

    (A) The pressure is in equilibrium with the gas.
    (B) The pressure is less than 1 atm.
    (C) The liquid is in equilibrium with the solid.
    (D) The pressure is greater than 1 atm.

158. Ethanol has a normal boiling point of 78°C. Ethanol boils at 78°C under which of the following conditions?

    (A) When the pressure is 1 atmosphere
    (B) At any pressure
    (C) When its molar volume is 22.4 L
    (D) At STP

159. Even though diamond and graphite both consist of carbon, diamond is one of the hardest substances known, while graphite is one of the softest. Diamond is a three-dimensional network of sp^3-hybridized carbon atoms with each carbon atom bonded to four others. Graphite, on the other hand, has a planar network of sp^2-hybridized carbon atoms, each bonded to three others in the plane through a resonating network. An explanation of why graphite is softer than diamond is that:

    (A) Graphite is resonance stabilized
    (B) sp^2 bonds are weaker than sp^3 bonds
    (C) Graphite has only weak London dispersion forces between the planes of carbon atoms instead of strong covalent bonds
    (D) The planes in graphite form a less compact structure than those in diamond

160. The molar enthalpy of sublimation is the amount of energy necessary to sublime a mole of a substance. Which of the following will have the highest enthalpy of sublimation?

    (A) Methanol
    (B) Formaldehyde
    (C) Ethane
    (D) Hydrogen chloride

161. Ozone forms in the stratosphere. This ozone helps protect the surface of the Earth from harmful ultraviolet radiation. The pressure in the stratosphere where the ozone is forming is about 8 torr. Which of the following statements is true?

    (A) The partial pressure of ozone in the stratosphere is less than 8 torr.
    (B) The partial pressure of ozone in the stratosphere is 8 torr.
    (C) The partial pressure of ozone in the stratosphere is greater than 8 torr.
    (D) The energy of the collisions of the gas molecules is lower at lower pressures.

162. All of these processes release energy to the surroundings *except*:

    (A) Freezing
    (B) Deposition
    (C) Condensation
    (D) Sublimation

163. A scientist needs to know the heat capacity of a solid for use in a calorimeter. If she has only the specific heat, what must she do?

    (A) Divide the specific heat by the molar mass of the solid.
    (B) Divide the specific heat by the mass.
    (C) Multiply the specific heat by the mass.
    (D) Multiply the specific heat by the molar mass of the solid.

164. The following is a generalized heating curve showing the transition of a substance from the solid state to the gaseous state. This diagram shows the temperature of the sample versus the heat added:

Heat added ⟶

Calculations concerning the "slanting" parts of the graph require information on which of the following?

(A) Heats of fusion
(B) Heats of vaporization
(C) Specific heats
(D) Heats of sublimation

165. The following is a generalized heating curve showing the transition of a substance from the solid state to the gaseous state. This diagram shows the temperature of the sample versus the heat added.

Heat added ⟶

When calculating the energy necessary to heat a sample from the melting point to the boiling point, it is necessary to know the amount of sample, the specific heat of the liquid, and:

(A) T_2 and T_3
(B) T_1 and T_3
(C) T_1 and T_4
(D) T_2 and T_4

166. How do the heats of fusion for the following substances compare?

NaCl   CCl$_4$ H$_2$O   Al$_2$O$_3$

(A) Al$_2$O$_3$ > NaCl > H$_2$O > CCl$_4$
(B) NaCl > CCl$_4$ > H$_2$O > Al$_2$O$_3$
(C) H$_2$O > CCl$_4$ > Al$_2$O$_3$ > NaCl
(D) CCl$_4$ > H$_2$O > NaCl > Al$_2$O$_3$

167. A sodium chloride solution is boiled. The predominant species in the vapor phase is/are:

(A) H$_2$O(g) and NaCl(g)
(B) H$_2$(g) and O$_2$(g)
(C) H$_2$O(g)
(D) H$_2$(g), O$_2$(g), and NaCl(g)

168. Even though diamond and graphite both consist of carbon, diamond is the hardest substance known, while graphite is one of the softest. However, the heat of sublimation of both substances is very high. Why is this so?

(A) In both cases, it is necessary to break strong covalent bonds.
(B) In both cases, it is necessary to break strong ionic bonds.
(C) In both cases, it is necessary to break the unusually strong London dispersion forces.
(D) In both cases, it is necessary to break strong metallic bonds.

169. Based upon the following thermochemical equations:

$$2\ H_2(g) + O_2(g) \rightarrow 2\ H_2O(l) \qquad \Delta H = -570\ kJ$$
$$N_2O_5(g) + H_2O(l) \rightarrow 2\ HNO_3(l) \qquad \Delta H = -80\ kJ$$
$$N_2(g) + 3\ O_2(g) + H_2(g) \rightarrow 2\ HNO_3(l) \qquad \Delta H = -350\ kJ$$

What is the value of ΔH for the following reaction?

$$2\ N_2(g) + 5\ O_2(g) \rightarrow 2\ N_2O_5(g)$$

(A) −30 kJ
(B) −3,390 kJ
(C) 30 kJ
(D) 3,390 kJ

170. Determine the specific heat of a metal sample if a 50.0 g sample absorbs 300.0 J of heat while the temperature rises from 25.0°C to 85.0°C.

(A) 0.020 J/g·K
(B) 0.100 J/g·K
(C) 0.0100 J/g·K
(D) 0.200 J/g·K

171. A 0.46 g sample of ethanol, C_2H_5OH, was placed in a bomb calorimeter and ignited according to the following equation:

$$C_2H_5OH(l) + 3 O_2(g) \rightarrow 2 CO_2(g) + 3 H_2O(l)$$

The total heat capacity of the calorimeter was 1.2 kJ/°C, and the reaction increased the temperature of the system from 22.50°C to 32.50°C. Determine the heat of reaction in kilojoules per mole of ethanol.

(A) -1.2×10^3 kJ
(B) $+1.2 \times 10^3$ kJ
(C) $+2.4 \times 10^3$ kJ
(D) -3.0×10^5 kJ

172. When an 8.6 g sample of sodium nitrate, $NaNO_3$, dissolved in 150.0 g of water in a coffee-cup calorimeter, the temperature changed from 25.3°C to 22.3°C. Calculate the enthalpy change, in kJ/mol $NaNO_3$, for sodium nitrate dissolving in water. Assume that the specific heat of the solution is the same as that of pure water (4.2 J/g·K).

(A) −20 kJ/mol
(B) +20 kJ/mol
(C) $+1.8 \times 10^3$ kJ/mol
(D) -1.8×10^3 kJ/mol

173. Hydrogen iodide, HI, melts at −50°C and boils at −35°C. For this compound, the enthalpy of fusion is 2.9 kJ/mol, and the enthalpy of vaporization is 44 kJ/mol. The molar specific heats for the liquid and gaseous phases are 44 J/mol·K and 25 J/mol·K, respectively. How much energy is necessary to convert 260 g of solid HI at the melting point to the vapor phase at 0.0°C?

(A) +100 kJ
(B) +140 kJ
(C) +3,300 kJ
(D) +1,000 kJ

174. Potassium metal reacts with water according to the following equation:

$$2\ K(s) + 2\ H_2O(l) \rightarrow 2\ KOH(aq) + H_2(g) \qquad \Delta H° = -130\ kJ$$

If the standard heat of formation for $H_2O(l)$ is -286 kJ/mol, what is the standard heat of formation for $KOH(aq)$?

   (A) +350 kJ/mol
   (B) −700 kJ/mol
   (C) −350 kJ/mol
   (D) Impossible to determine

175. Why does methanol have a higher molar heat of vaporization than methyl chloride?

   (A) The molecular weight of methyl chloride is higher than that of methanol.
   (B) The H–O bonds in methanol are stronger than any of the bonds in methyl chloride.
   (C) Dipole-dipole forces in methyl chloride are weaker than the intermolecular forces in methanol.
   (D) Chlorine causes methyl chloride to be more polar than methanol.

176. The heat of vaporization of acetone is 0.500 kJ/g, and the boiling point of acetone is 56°C. How much heat is necessary to vaporize 100.0 g of acetone at 329 K?

   (A) 1.36×10^4 kJ
   (B) 50.0 kJ
   (C) 100.0 kJ
   (D) Not possible to determine

177. During an experiment, a graduate student adds 100 J of energy to a sample of ethanol at a temperature below the boiling point. The addition of energy causes no change in temperature. Which of the following could explain this observation?

   (A) More energy is necessary to change the temperature.
   (B) The sample is at its melting point.
   (C) The bond energies are all greater than 100 J, so no bonds could be broken.
   (D) The energy dissipates as friction.

178. The boiling point of ethanol is 78°C. At the boiling point, there is a phase change from liquid to gas. The enthalpy change for this process is:

    (A) Positive because *intra*molecular forces are being overcome
    (B) Positive because *inter*molecular forces are being overcome
    (C) Negative because *inter*molecular forces are being overcome
    (D) Negative because *intra*molecular forces are being overcome

179. The following standard enthalpies of formation are reported for sulfur trioxide (SO_3):

$SO_3(s)$	−454 kJ/mol
$SO_3(l)$	−441 kJ/mol
$SO_3(g)$	−396 kJ/mol

    What is the standard heat of deposition for sulfur trioxide?

    (A) +58 kJ/mol
    (B) −58 kJ/mol
    (C) −13 kJ/mol
    (D) +13 kJ/mol

180. The following is a generalized heating curve showing the transition of a substance from the solid state to the gaseous state. This diagram shows the temperature of the sample versus the heat added:

Heat added ⟶

    In most cases, the second horizontal region is longer than the first (at a constant heating rate). Why is the second horizontal region longer?

    (A) The heat of vaporization is greater than the heat of fusion.
    (B) The heat of fusion is greater than the heat of vaporization.
    (C) The heat of vaporization is greater than the heat of sublimation.
    (D) The heat of fusion is greater than the heat of condensation.

181. A 25 g sample of a substance was dissolved in 175 g of water. The temperature of the resultant solution was 10.0°C higher than that of the original water. Assuming the specific heat of the solution was the same as that of pure water (4.2 J/g°C), what was the enthalpy change for the process?

    (A) −8.4 kJ
    (B) +8.4 kJ
    (C) −1.1 kJ
    (D) +7.4 kJ

182. At the critical point on a phase diagram, the:

    (A) Gas ceases to exist
    (B) Liquid ceases to exist
    (C) Liquid vaporizes
    (D) Liquid and gas phases are indistinguishable

CHAPTER 5

Gases

Questions 183–187 refer to the passage below.

Gases are those substances found in the gaseous state at room temperature, as opposed to vapors. Vapors are the gaseous state of substances that commonly exist in the liquid or solid state at room temperature. At the macroscopic level, gases are highly compressible, and unlike solids and liquids, they expand to fill the volume of their container. At the microscopic level, gases are normally low atomic or molecular mass atoms or molecules, and there is a relatively large distance between the gas particles (which explains their compressibility). Solids and liquids have much less space between the particles and thus are called **condensed phases**. The gas particles possess a relatively large amount of kinetic energy, enough to overcome the attractive forces between the particles (**intermolecular forces**), so that the particles are relatively independent of each other. Finally, gases always form homogeneous mixtures when mixed.

While individual gases have unique properties associated with them, all gases behave in a similar fashion with respect to temperature, pressure, volume, and amount (moles). Relationships between pairs of these conditions can be determined if the other two are held constant. These relationships were discovered between the mid-1600s and the mid-1800s and are named after the individuals who discovered them. Taken together, they are the gas laws.

183. A student builds the following apparatus:

In an experiment, the student began with a 0.50 mole sample of argon gas at 0°C and 1.00 atm. The initial volume was 11 L. The sample was heated at constant pressure to 273°C and the final volume determined. What was the final volume?

(A) 11 L
(B) 273×11 L
(C) 22 L
(D) 44 L

184. A student builds the apparatus shown in question 183. In an experiment, the student began with a 0.50 mole sample of argon gas at 0°C and 1.00 atm, and the piston was locked in place. The sample was then heated to 273°C and the final volume determined. How do the final pressure (P_f) and volume (V_f) relate to the initial pressure (P_i) and volume (V_i)?

(A) $P_f = 2\,P_i$ and $V_f = 2\,V_i$
(B) $P_f = P_i$ and $V_f = V_i$
(C) $P_f = 2\,P_i$ and $V_f = V_i$
(D) $P_f = P_i$ and $V_f = 2\,V_i$

185. A student builds the apparatus shown in question 183. In an experiment, the student began with a 0.50 mole sample of argon gas at 0°C and 1.00 atm, and the piston was locked in place. The sample was then heated to 273°C, and the initial pressure (P_i) and the volume (V_i) were measured. At this point, the piston is unlocked, and a heater/cooler keeps the temperature constant. How do the final pressure (P_f) and volume (V_f) relate to the initial pressure (P_i) and volume (V_i)?

(A) $P_f < P_i$ and $V_f > V_i$
(B) $P_f < P_i$ and $V_f < V_i$
(C) $P_f > P_i$ and $V_f > V_i$
(D) $P_f = P_i$ and $V_f > V_i$

186. A sample of a gas is confined in a cylinder with a moving piston. If the diameter of the cylinder is halved and the force on the gas is also halved, how will the volume change?

(A) No change
(B) Increase
(C) Decrease
(D) Unknowable without knowing the temperature

187. A student builds the apparatus shown here:

In an experiment, the student began with a 0.50 mole sample of argon gas at 0°C and 1.00 atm. The sample was heated at constant pressure to 273°C and the final volume determined. What was the initial volume?

(A) $1.1 \times 10^{-2} \text{ m}^3$
(B) $2.2 \times 10^{-2} \text{ m}^3$
(C) $5.6 \times 10^{-3} \text{ m}^3$
(D) $4.4 \times 10^{-2} \text{ m}^3$

188. A mixture of gases has a total pressure of 1,000 torr. The mixture contains 25% methane, 40% oxygen, and 35% hydrogen. What is the partial pressure of hydrogen?

(A) 400 torr
(B) 35 torr
(C) Impossible to determine
(D) 350 torr

189. A steel cylinder contains 1 mole of an ideal gas. At constant pressure, increasing the moles of gas will cause:

(A) The temperature to increase
(B) The temperature to decrease
(C) The temperature to not change
(D) An undetermined change in the temperature unless the actual volume is known

190. Air is primarily nitrogen and oxygen, with small amounts of other gases. Which of the following statements is true concerning the nitrogen and oxygen in air at room temperature?

    (A) Oxygen is heavier, so its molecules must move faster to compensate.
    (B) Oxygen is heavier, so it has a higher kinetic energy.
    (C) Both gases have the same kinetic energy.
    (D) Kinetic energy determinations are not possible due to interference by the other gases.

191. Many shock absorbers rely on air to function. Which of the following is the *best* reason why water could not replace the air in a shock absorber?

    (A) Water freezes as the pressure increases.
    (B) Water freezes at 32°F.
    (C) Water is denser than air.
    (D) Water is nearly incompressible.

192. How will the absolute temperature change when halving the pressure on a sample of an ideal gas at constant volume?

    (A) It remains constant.
    (B) It doubles.
    (C) It halves.
    (D) It depends on the number of moles.

193. A number of equations are applicable to the behavior of gases. The basis used for the explanation of the diffusion of gases is:

    (A) $\left(P + \dfrac{an^2}{v^2}\right)(V - nb) = nRT$

    (B) $P_1V_1 = P_2V_2$
    (C) $PV = nRT$
    (D) $KE = 1/2\ mv^2$

194. The following apparatus was used to examine gas phase reactions. The two compounds react to form a nonvolatile compound, which is identified by the formation of a cloud.

In one experiment, the volatile base was dimethylamine, $(CH_3)_2NH$, and the cloud formed very near the center. The identity of the other reactant might be:

(A) HI
(B) HF
(C) HCOOH
(D) HSCN

195. Which of the following gases is most likely to be ideal under normal conditions?

(A) Xe
(B) He
(C) HF
(D) C_2H_6

196. A sample of a gas mixture contains helium, neon, and argon. The partial pressures of these gases are He = 0.50 atm, Ne = 76 torr, and Ar = 10 kPa. What is the total pressure of the mixture?

(A) 0.70 atm
(B) 10 kPa
(C) 0.50 atm
(D) 1.0 atm

197. A mixture of gases made of 32.0 g of CH_4, 60.0 g of C_2H_6, and 116 g of C_4H_{10} was contained in a flask. The total pressure of the gases was 2.40 atm. Determine the partial pressure of each gas, in atmospheres.

    (A) $CH_4 = 0.40$ atm, $C_2H_6 = 0.70$ atm, and $C_4H_{10} = 1.30$ atm
    (B) $CH_4 = 0.80$ atm, $C_2H_6 = 0.80$ atm, and $C_4H_{10} = 0.80$ atm
    (C) $CH_4 = 0.30$ atm, $C_2H_6 = 0.50$ atm, and $C_4H_{10} = 1.60$ atm
    (D) Impossible to determine without knowing the volume of the flask

198. A sample of a gas contains 1.0 mole of a gas at 1.0 atmosphere and 0°C. The density of the gas is 5.8 g/L. What might be the identity of the gas?

    (A) Xe
    (B) CO_2
    (C) CH_4
    (D) H_2

199. A sample of a gas is held at constant pressure. Initially, the gas occupied 1.0 L at a temperature of 127°C. The temperature was changed until the volume became 500 mL. What was the temperature at the new volume?

    (A) 64°C
    (B) −73°C
    (C) 73°C
    (D) −13°C

200. A student has two balloons. Both balloons have the same volume, pressure, and temperature. One of the balloons contains hydrogen gas, and the other balloon contains oxygen gas. Which of the following will be greater for the oxygen balloon?

    (A) Moles
    (B) Average kinetic energy
    (C) Density
    (D) The properties are all the same.

201. A student has two balloons. Both balloons have the same volume, pressure, and temperature. One of the balloons contains hydrogen gas, and the other balloon contains oxygen gas. How do the speeds and average kinetic energies of molecules in each balloon compare?

(A) The hydrogen molecules are moving faster, but they have the same average kinetic energy as the oxygen molecules.

(B) The hydrogen molecules are moving slower, but they have the same average kinetic energy as the oxygen molecules.

(C) The hydrogen molecules are moving faster, and they have a lower average kinetic energy than the oxygen molecules.

(D) The hydrogen molecules are moving faster, and they have a higher average kinetic energy than the oxygen molecules.

202. The rate of effusion for an unknown gas was found to be half that of methane gas under the same conditions. Which of the following might be the unknown gas?

(A) O_2
(B) He
(C) CO_2
(D) SO_2

203. The van der Waals constants for three gases are given in the following table:

Gas	a (L^2 atm/mol^2)	b (L/mol)
CO_2	3.658	0.04286
SO_2	6.865	0.05679
O_2	1.382	0.03186

Which of the gases is the least ideal?

(A) O_2
(B) CO_2
(C) SO_2
(D) All are equally far from ideal.

204. Gas samples approach ideal behavior at low pressure and high temperature. Which postulate of kinetic molecular theory is a problem at high pressure?

(A) A gas is composed of particles called atoms or molecules.
(B) The volumes of the gas particles are negligible.
(C) The average kinetic energy of the molecules depends on the absolute temperature.
(D) The particles are constantly moving.

205. Gas samples approach ideal behavior at low pressure and high tempera-ture. Which postulate of kinetic molecular theory is a problem at low temperature?

    (A) There are no attractive or repulsive interactions between the gas particles.
    (B) The volumes of the gas particles are negligible.
    (C) The average kinetic energy of the molecules depends on the absolute temperature.
    (D) The particles are constantly moving.

206. Two gas samples initially have the same volume (500 mL), temperature (27°C), and pressure (1 atm). The pressure on one sample is halved, while the absolute temperature of the other sample is doubled. What are the rela-tive volumes of the two samples after the changes?

    (A) The lower-pressure sample is larger.
    (B) The heated sample is larger.
    (C) They are both the same.
    (D) The lower-pressure sample is smaller.

207. A student does an experiment examining the properties of water vapor. During each step of the experiment, she finds that the pressure of the water vapor is a little lower than expected from $PV = nRT$. Why is the pressure low?

    (A) There is a strong attraction between the water molecules.
    (B) The water molecules are not linear in the gas phase.
    (C) The gaseous water molecules are nearly ideal.
    (D) There was some liquid water present in the sample.

208. The volume of 56 g of N_2 at STP is:

    (A) 56.0 L
    (B) 28.0 L
    (C) 44.8 L
    (D) 22.4 L

209. A gaseous mixture containing helium, neon, and argon has a total pressure of 176 torr. The partial pressure of helium is 63 torr, and the partial pres-sure of neon is 25 torr. What percent of the total pressure is due to argon?

    (A) 33%
    (B) Impossible to determine
    (C) 50%
    (D) 25%

210. A sample contains two gases, A and B. The velocity of gas B is twice that of gas A. What are the possible identities of gases A and B?

    (A) Kr and He
    (B) O_2 and N_2
    (C) H_2 and Xe
    (D) Kr and Ne

211. A latex balloon contains a mixture of helium contaminated with air. The helium causes the balloon to rise. The next day, the balloon does not rise to the same degree as previously. Why?

    (A) The light helium rapidly effuses out of the balloon.
    (B) Heavier air molecules effuse into the balloon.
    (C) The helium molecules slow down overnight, which lowers the temperature and causes the balloon to contract.
    (D) Water vapor effuses into the balloon and condenses.

212. The following apparatus was used to examine gas phase reactions. The two compounds react to form a nonvolatile compound, which is identified by the formation of a cloud.

In one experiment, the volatile base was methylamine, CH_3NH_2, and a series of HX materials were used, which yielded the following data:

Sample	Position of cloud (relative to 0)
HX1	10 cm
HX2	17 cm
HX3	23 cm

According to these results:

    (A) HX1 has the highest molecular weight
    (B) HX1 has the highest vapor pressure
    (C) HX1 is the strongest acid
    (D) HX1 has the lowest activation energy

213. Which of the following is *not* based on an assumption of kinetic molecular theory?

    (A) The intermolecular forces between the molecules of all gases are negligible.
    (B) At a given temperature, the average kinetic energy of all gas molecules is the same.
    (C) The volume of all gas molecules is negligible compared to the volume of the container.
    (D) At a given temperature, the average velocity of all gas molecules is the same.

214. A sample of a gas mixture contains helium, neon, and argon. The partial pressures of these gases are He = 0.50 atm, Ne = 0.20 atm, and Ar = 0.30 atm. What are the mole fractions of each gas?

    (A) He = 50, Ne = 20, Ar = 30
    (B) He = 1.0, Ne = 0.40, Ar = 0.60
    (C) He = 0.50, Ne = 0.20, Ar = 0.30
    (D) Impossible to determine without knowing either the volume or the temperature

215. A sample of a gas contains 2.0 mole of gas at 2.0 atmosphere pressure and 249°C. What is the volume of the gas?

    (A) 20.4 L
    (B) 22.4 L
    (C) 42.8 L
    (D) 10.2 L

216. A sample of a volatile liquid is placed in an 80 mL container and volatilized. The vapor exerts a pressure of 3.0 atm at a temperature of 27°C. The mass of the vapor is 1.2 g. What might be the identity of the volatile liquid?

    (A) CCl_4
    (B) CH_3OH
    (C) C_8H_{18}
    (D) C_4H_{10}

217. A student used the following reaction to generate hydrogen gas:

$$Zn(s) + H_2SO_4(aq) \rightarrow ZnSO_4(aq) + H_2(g)$$

The student collected the gas by water displacement. The complete reaction of 0.010 mole of zinc with excess sulfuric acid generated 224 mL of gas at 0°C. The student used the ideal gas equation to calculate the pressure of the gas. The calculated pressure was 1.0 atm. However, when the student measured the pressure, he found that it was greater than calculated. Which of the following best explains the discrepancy?

(A) The student did not include the vapor pressure of water.
(B) The student forgot to account for the volatility of sulfuric acid.
(C) The kinetic energy of the hydrogen molecules is higher at the freezing point of water.
(D) The student did not consider the formation of ice in the system.

CHAPTER 6

Solutions

Questions 218–225 refer to the following passage.

Colligative properties are solution properties that depend only upon the number of solute particles present. The presence of a solute will alter the freezing and boiling points of a solvent because of the lowering of the vapor pressure of the solvent by the solute. The freezing point of a solution will be lower than the freezing point of the pure solvent, and the boiling point of a solution will be higher than the boiling point of the pure solvent. The change in the freezing or boiling point is represented by DT. The decrease in the freezing point is the **freezing point depression**. The increase in the boiling point is the **boiling point elevation**. The means of determining the freezing point depression and the boiling point elevation involve similar calculations.

Freezing point depression and boiling point elevation are colligative properties. Thus, the number of particles present in the solution is the key. Table salt is a strong electrolyte. Ideally, the concentration of NaCl in the solution is not 1.00 m; it is 0.00 m. The solution is 1.00 m in Na^+(aq) and 1.00 m Cl^-(aq), for a total of 2.00 m. Since the concentration of particles in the NaCl solution is 2.00 m, the freezing point depression is twice the depression of a 1.00 m sugar solution.

Remember to convert strong electrolytes to the individual ions and add the separate concentrations together before using either of the DT equations. This is one way of determining the change in temperature. An alternative method is to use a van't Hoff factor, i. The **van't Hoff factor** is the number of ions produced by an electrolyte in solution. The value of the van't Hoff factor for sodium chloride is 2.

218. The freezing point of a solution is depressed by an amount determined by the relationship $\Delta T = iK_f m$, where i is the van't Hoff factor (the number of mole particles produced by each mole of solute), K_f is the solvent-specific freezing point depression constant, and m is the molality of the solution. A solution is found to have $\Delta T = -0.80°C$. Before the researcher can repeat the experiment, he spills half of the solution. However, an attempt is made to determine the freezing point of the remaining solution. The freezing point of the remaining solution is:

(A) −0.40°C
(B) −0.80°C
(C) −1.60°C
(D) Impossible to determine

219. The van't Hoff factor, i, is a term used to express the number of particles produced by an electrolyte in solution. The value of this factor may vary due to interactions between the particles. Under which of the following conditions would the van't Hoff factor be least likely to vary from the ideal value?

(A) Low concentrations
(B) High concentrations
(C) High temperatures
(D) Low pressures

220. A scientist measures the freezing point depression of a 0.1 m sodium nitrate solution. He uses the ideal van't Hoff factor for the calculation of the ideal freezing point depression. The observed freezing point is:

(A) Not measurable because the solution did not freeze
(B) Lower than the theoretical
(C) Less than the change in the boiling point
(D) Higher than the theoretical

221. The van't Hoff factor, i, is a term used to express the number of particles produced by an electrolyte in solution. The value of this factor may vary due to interactions between the particles. There are a number of methods available to determine the variation in the van't Hoff factor. Which of the following experiments would *not* lead to a determination of this variation?

(A) A density measurement of the solution
(B) A freezing point measurement of the solution
(C) A boiling point measurement of the solution
(D) An osmotic pressure measurement of the solution

222. If the boiling point constant, K_b, for water is about 0.5°C/mol, what is the approximate boiling point of a 1 mol solution of potassium phosphate?

(A) 100.5°C
(B) 99.5°C
(C) 98.0°C
(D) 102.0°C

223. Two solutions were prepared. One solution was 1.0 M $NaNO_3$(aq), and the other was 1.0 M $Ca(NO_3)_2$(aq). Which solution will freeze at the lower temperature?

(A) Both will freeze at the same temperature because the molarities are equal.
(B) $Ca(NO_3)_2$ will because there are more ions present in the solution.
(C) $NaNO_3$ will because there are more ions present in the solution.
(D) $Ca(NO_3)_2$ will because its molar mass is greater.

224. The van't Hoff factor, i, is a term used to express the number of particles produced by an electrolyte in solution. The value of this factor may vary due to interactions between the particles. The interactions of the particles in a $CuSO_4$ solution are greater than the interactions of the particles in a KCl solution. Why is this so?

(A) Transition metal ions can form complexes.
(B) Sulfate ions are larger than chloride ions.
(C) The charges of the ions are greater.
(D) The sulfate ion undergoes significant hydrolysis.

225. The van't Hoff factor, i, is a term used to express the number of particles produced by an electrolyte in solution. What is the van't Hoff factor of $(NH_4)_2SO_4$?

(A) 15
(B) 3
(C) 1
(D) 2

226. A sample of hydrogen gas was collected over water. Only part of the pressure of the collected gas was due to hydrogen. The remaining pressure was due to:

(A) Water vapor
(B) Oxygen gas
(C) Liquid hydrogen
(D) Inert gases

227. Why does NaCN produce a greater freezing point depression than HCN?

   (A) Only NaCN dissociates completely in solution.
   (B) Only HCN dissociates completely in solution.
   (C) The heat of solution for NaCN is more endothermic than that of HCN.
   (D) The heat of solution for HCN is more endothermic than that of NaCN.

228. Water has a vapor pressure of 32 torr at 30°C. What will be the vapor pressure if 10.0 g of magnesium chloride is added to 100.0 mL of water at 30°C?

   (A) It will be lower.
   (B) It will be higher.
   (C) It will be the same.
   (D) It will be the average of the vapor pressure of the two substances.

229. Sulfur trioxide gas (SO_3) is most soluble in which of the following solutions?

   (A) 1.0 M NaOH
   (B) 1.0 M Na_2SO_3
   (C) 1.0 M Na_2SO_4
   (D) 1.0 M HCl

230. Two grams of an unknown salt was added to 100.0 mL of pure water. The salt completely dissolved. Which of the following might be the salt?

   (A) Ag_2S
   (B) $CaCO_3$
   (C) $AlPO_4$
   (D) $Ca(C_2H_3O_2)_2$

231. The vapor pressure of water does *not* depend on:

   (A) The external pressure
   (B) The temperature
   (C) Hydrogen bonding
   (D) The presence of solutes

232. Ammonia gas is very soluble in water. Which of the following statements is true concerning the solution formed?

(A) The ammonia completely ionizes.
(B) The boiling point is higher than that of pure water.
(C) The greater the concentration of ammonia, the lower the pH.
(D) None of the above

233. Why is decanoic acid ($C_9H_{19}COOH$) more soluble in a basic solution than in pure water?

(A) The acid reacts with the base to produce a hydrophilic ion.
(B) The acid reacts with the base to produce a hydrophobic ion.
(C) Weak ion-dipole forces replace the hydrogen bonds present in the pure acid.
(D) A basic solution contains hydrated ions that are not present in pure water.

234. Which of the following solutions will boil at the highest temperature?

(A) 1.0 M HNO_3
(B) 1.0 M NaCl
(C) 2.0 M CH_3CH_2OH
(D) 1.0 M $CaCl_2$

235. Salts such as sodium chloride and calcium chloride lower the freezing points of aqueous solutions. The reason this occurs is that:

(A) It is necessary to cool the solution further to compensate for the heat of solution of the salts
(B) The dissolved salts increase the density of the solution
(C) The ions from the dissolved salts interfere with the formation of ice crystals
(D) The dissolved salts lower the vapor pressure of the solution

236. In osmosis, a semipermeable membrane separates two solutions. Which of the following statements is true concerning osmosis?

(A) The osmotic pressure of a 1.0 m sodium chloride solution is the same as that of a 1.0 m acetic acid solution.
(B) The solvent does not flow if the concentrations are the same in both solutions.
(C) The solvent always flows into the solution of higher concentration.
(D) Osmotic pressure forces the solvent through the membrane.

237. Ion-exchange resins, such as water softeners, are capable of removing ions from solution. During the removal, ions in the solution displace ions already on the resin. The exchange occurs because the ions in the solution have a greater affinity for the resin than the ions already present. Most water softeners begin with sodium ions on the resin. A sodium ion water softener would be least effective in removing which of the following ions from solution?

    (A) Li^+
    (B) K^+
    (C) Mg^{2+}
    (D) Ca^{2+}

238. Reverse osmosis is capable of removing ions from solution through the application of a pressure greater than the osmotic pressure. Prior to the application of pressure, a reducing agent is often used to treat the water. Which of the following ions will the reducing agent be most likely to affect?

    (A) Hg^{2+}
    (B) Na^+
    (C) Ca^{2+}
    (D) Al^{3+}

239. It is possible to purify water by distillation. However, purification by distillation is less effective if the contaminant is volatile. Distillation would work *best* for which of the following contaminants?

    (A) $C_2H_2Br_2$
    (B) CH_2Cl_2
    (C) NCl_3
    (D) $HgCl_2$

240. The vapor pressure of water is about 0.030 atm at 25°C. What is the vapor pressure of a solution made by adding 5.0 moles of antifreeze (ethylene glycol) to 1.0 L of water?

    (A) 0.040 atm
    (B) 0.027 atm
    (C) 0.015 atm
    (D) 0.0080 atm

241. Four 1 M test solutions are cooled at equal rates. The identities of the test solutions are C_2H_5OH, $NaNO_3$, Na_3PO_4, and Na_2S. Which would be the first solution to freeze?

    (A) C_2H_5OH
    (B) $NaNO_3$
    (C) Na_3PO_4
    (D) Na_2S

242. A precipitate will form when mixing a barium chloride solution with each of the following solutions: $Na_2C_2O_4$, $NaIO_3$, Na_2CO_3, and Na_3PO_4. Which solution will give a precipitate with the strongest ionic bonding?

    (A) Na_3PO_4
    (B) Na_2CO_3
    (C) $NaIO_3$
    (D) $Na_2C_2O_4$

243. Mixing 50.00 mL of 4.00 M NaOH with 50.00 mL of 4.00 M HNO_3 in a calorimeter resulted in a 52°C temperature increase. The freezing point of the final solution would be:

    (A) Higher than that of either of the original solutions
    (B) Lower than that of either of the original solutions
    (C) Equal to that of either of the original solutions
    (D) The average of the freezing points of the original solutions

244. The experimental measurement of the osmotic pressure of a solution containing $Pt(NH_3)_4Cl_2$ indicated that the compound ionized into three ions in aqueous solution. What is the most likely geometry of the cationic complex?

    (A) Linear
    (B) Tetrahedral
    (C) Octahedral
    (D) Square planar

245. Which of the following is more soluble in 1.0 M H_2SO_4 than in pure water?

    (A) $Na_2S(s)$
    (B) $CaI_2(s)$
    (C) $Ca(NO_3)_2(s)$
    (D) $KBr(s)$

246. Which of the following is the most soluble in water?

(A) $CH_2(OH)CH_2CH_2CH(OH)CH_2CH_2CH(OH)$-
$CH_2CH_2CH_2CH_2OH$
(B) $CH_2(OH)CH_2CH_2CH(OH)CH_2CH_2CH_2CH_2CH_2CH_2OH$
(C) $CH_2(OH)CH_2CH_2CH_2CH_2CH_2CH_2CH_2CH_2CH_2OH$
(D) $CH_3CH_2CH_2CH_2CH_2CH_2CH_2CH_2CH_2CH_2CH_2OH$

247. Which of the following properties does *not* depend on the number of particles present in solution?

(A) Vapor pressure
(B) Density
(C) Freezing point
(D) Boiling point

248. $NH_3(g)$ and $HCl(g)$ are more soluble in water than $O_2(g)$ is. Why?

(A) Elements (O_2) are less soluble than compounds (HCl and NH_3).
(B) O_2 has a double bond, which is less reactive than the single bonds in NH_3 and HCl.
(C) NH_3 and HCl are polar and O_2 is not.
(D) O_2 causes combustion, but NH_3 and HCl do not support combustion.

249. Which of the following solutions will behave closest to an ideal solution?

(A) 0.01 M $CaCl_2$
(B) 0.01 M $NaCl$
(C) 0.01 M Na_2CO_3
(D) 0.01 M $Al_2(SO_4)_3$

250. At a given temperature, the vapor pressure of Liquid A is greater than that of Liquid B. All of the following are true about Liquid A *except*:

(A) Liquid A has a higher boiling point.
(B) Liquid A has a lower boiling point.
(C) Liquid A has weaker intermolecular forces than Liquid B.
(D) The freezing point of Liquid A is lower than that of Liquid B.

251. An important factor leading to oxygen/carbon dioxide exchange in the lungs is:

(A) A concentration gradient
(B) A temperature difference
(C) The volume change as the lungs expand
(D) The lower pH of blood in the lungs

252. Which of the following solutions will have the highest boiling point?

(A) 1.0 m NaNO$_3$
(B) 1.0 m CaCl$_2$
(C) 1.0 m Al(NO$_3$)$_3$
(D) 2.0 m C$_2$H$_5$OH

253. The first compound isolated during the fractional distillation of a mixture of PF$_3$, AsF$_3$, and SbF$_3$ would be:

(A) AsF$_3$
(B) PF$_3$
(C) SbF$_3$
(D) Fractional distillation will not work on this mixture.

254. Four test solutions are used to test an unknown. In each case, the concentration of the test solution and the unknown solution was 1 M. The test solutions are NaCl, NaNO$_3$, NaI, and Na$_2$S. The results of mixing each of the test solutions with an unknown were NaCl—white precipitate, NaNO$_3$—no reaction, NaI—yellow precipitate, and Na$_2$S—black precipitate. The unknown might be:

(A) AgNO$_3$
(B) KBr
(C) Ca(NO$_2$)$_2$
(D) Al$_2$(SO$_4$)$_3$

255. An aqueous sodium carbonate solution will react with many aqueous transition metal ion solutions to form a precipitate. Which of the following solutions would be the most effective at redissolving the precipitate?

(A) 1 M C$_2$H$_5$OH
(B) 1 M NaOH
(C) 1 M HNO$_3$
(D) 1 M Na$_2$CO$_3$

CHAPTER 7

Kinetics

Questions 256–262 refer to the following passage.

The rate or speed of a chemical reaction depends upon several factors in addition to the identities of the reactants. Some of these factors include the concentration of reactants, their physical state, and the temperature.

In general, the greater the concentration of the reactants, the faster will be the reaction. Atoms or molecules must come into contact in order to react. The greater the concentration of a reactant, the more likely it is for one reactant to come into contact with another reactant and react.

The physical state is also an important factor. For example, consider the following reaction:

$$Ca(s) + S(s) \rightarrow CaS(s)$$

If you place a block of calcium metal and a block of sulfur together on a table one inch apart, no reaction will occur. If you now move the blocks together, a reaction will begin along the plane of contact. However, once all the calcium and sulfur atoms along the plane of contact have reacted, the reaction will cease. We could increase the amount of reaction by grinding both reactants into a fine powder and mixing the powders. The increased surface area would allow many more atoms to come into contact and react. Therefore, the reaction would proceed at a somewhat faster rate than between the blocks.

Consider another reaction:

$$H_2(g) + Cl_2(g) \rightarrow 2\ HCl(g)$$

We would not have the same problem as we had with solid reactants. Both of these reactants are gases, and gases readily mix. The rapid mixing of these gases quickly allows the molecules to come into contact and react. Therefore, in general, gases react more rapidly than solids.

Temperature is an important factor. At higher temperatures, molecules are moving faster. The faster molecules move, the more likely they are to come into contact and react.

Studying the rates of chemical reactions allows us to develop relationships called rate laws. If we wish to know when a reaction is complete, we can use a rate law. This is the type of information a chemical engineer must know when designing a chemical plant. If we want to know when a reaction started, we can use a rate law. This is exactly what an archaeologist does when dating an ancient artifact. When we want to know when a patient needs another dose of a drug, we can use a rate law.

256. What happens to the rate of a reaction when the temperature increases by 30°C?

    (A) The reaction continues at the same rate, and the rate constant increases.

    (B) The rate of the reaction increases by a factor of 8.

    (C) The rate of the reaction increases by a factor of 2.

    (D) The reaction continues at the same rate, and the rate constant decreases.

257. The following table contains data obtained in the study of the following reaction:

$$A + B \rightarrow C$$

Experiment	$[A]_{initial}$	$[B]_{initial}$	Initial Rate (M/s)
1	0.1 M	0.1 M	4.0×10^{-2}
2	0.2 M	0.1 M	8.0×10^{-2}
3	0.2 M	0.2 M	1.6×10^{-1}

    What is the rate law for this reaction?

    (A) Rate $= k[C]/([A][B])$

    (B) Rate $= k[A][B]$

    (C) Rate $= k[A]^2[B]$

    (D) Rate $= k[A][B]^2$

258. A certain reaction follows the rate law Rate $= k[A]^2[B]$. By what factor will the rate of the reaction change if the concentration of A is halved and the concentration of B is doubled?

    (A) 1/4

    (B) 2

    (C) 4

    (D) 1/2

259. The reaction of iodide ion, I⁻, with hypochlorite ion, ClO⁻, will take place in basic solution. The overall reaction is:

$$I^-(aq) + ClO^-(aq) \rightarrow IO^-(aq) + Cl^-(aq)$$

The following data were obtained at a certain temperature:

Experiment	[I⁻] (M)	[ClO⁻] (M)	[OH⁻] (M)	Relative rate of IO⁻ formation (M/s)
1	0.10	0.40	0.10	2
2	0.20	0.40	0.10	4
3	0.10	0.20	0.10	1
4	0.10	0.60	0.10	8
5	0.10	0.40	0.05	4
6	0.10	0.40	0.20	1
7	0.10	0.60	0.20	1.5

What is the order of the reaction for each of the substances?

(A) $[I^-] = +1$ $[ClO^-] = +1$ $[OH^-] = -1$
(B) $[I^-] = +1$ $[ClO^-] = +1$ $[OH^-] = 0$
(C) $[I^-] = 0$ $[ClO^-] = -1$ $[OH^-] = +1$
(D) $[I^-] = -1$ $[ClO^-] = 0$ $[OH^-] = 0$

260. The reaction of iodide ion, I⁻, with hypochlorite ion, ClO⁻, takes place in basic solution. Referring to the same reaction and data as question 259, what is the role of the hydroxide ion?

(A) It is a catalyst.
(B) It is an inhibitor.
(C) It is a promoter.
(D) It is a scavenger.

261. The following reaction illustrates the decomposition of acetaldehyde:

$$CH_3CHO(g) \rightarrow CH_4(g) + CO(g)$$

This reaction is second order with a rate of 1.6×10^{-2} M s⁻¹ at a certain temperature. Calculate the rate constant at this temperature if the initial concentration of acetaldehyde is 0.40 M.

(A) 2.6×10^{-3} M⁻¹ s⁻¹
(B) 6.4×10^{-3} M⁻¹ s⁻¹
(C) 0.040 M⁻¹ s⁻¹
(D) 1.0×10^{-1} M⁻¹ s⁻¹

262. The experimental investigation of a given reaction generated the following data:

$$2 NO(g) + O_2(g) \rightarrow 2 NO_2(g)$$

Experiment	[NO], M	[O_2], M	Rate, $\Delta[NO_2]/\Delta t$
1	0.0150	0.0150	2.40×10^{-2}
2	0.0300	0.0300	1.92×10^{-1}
3	0.0300	0.0150	9.59×10^{-2}

What is the rate law for this reaction?

(A) Rate $= k[NO]^2[O_2]$
(B) Rate $= k[NO][O_2]$
(C) Rate $= k\left(\dfrac{[NO_2]^2}{[NO]^2[O_2]}\right)$
(D) Rate $= k[NO_2]^2$

263. The reaction of hydrogen with oxygen and the reaction of hydrogen with nitrogen are both exothermic [$\Delta H(O_2) = -241$ kJ/mol and $\Delta H(N_2) = -17$ kJ/mol]. However, the reaction of hydrogen with nitrogen is slower. Why?

(A) The activation energy is higher.
(B) The $\Delta H(N_2)$ is smaller.
(C) H_2O is more stable than NH_3.
(D) O_2 supports combustion.

264. For a particular chemical reaction, the rate constant is independent of:

(A) Activation energy
(B) Enthalpy change for the reaction
(C) Temperature
(D) Presence or absence of a catalyst

265. Sulfides react with acids to release hydrogen sulfide gas. The rate of the reaction is directly related to the hydrogen ion concentration. Which of the following will cause the release of hydrogen sulfide at the fastest rate?

(A) 0.5 M HCl
(B) 0.5 M HF
(C) 0.5 M H_2SO_3
(D) 0.5 M HNO_2

266. Nitrogen oxide catalyzes the destruction of ozone in the upper atmosphere. One suggested mechanism is:

$$2 [NO(g) + O_3(g) \rightarrow NO_2(g) + O_2(g)] \qquad \text{fast}$$
$$2 NO_2(g) \rightarrow N_2O_4(g) \qquad \text{slow}$$
$$N_2O_4(g) + hv \rightarrow 2 NO(g) + O_2(g) \qquad \text{fast}$$

Which of the following statements is true about the energy profile?

(A) The plot has only one peak, which depends on the first step.
(B) The plot has only one peak, which depends on the second step.
(C) The plot has three peaks, with the relative height of the central peak being the lowest.
(D) The plot has three peaks, with the relative height of the central peak being the highest.

267. The following reaction begins very slowly but rapidly speeds up as $Mn^{2+}(aq)$ forms:

$$3 H_2SO_4(aq) + 2 KMnO_4(aq) + 5 H_2C_2O_4(aq) \rightarrow$$
$$K_2SO_4(aq) + 2 MnSO_4(aq) + 10 CO_2(g) + 8 H_2O(l)$$

Which of the following would change the value of the rate constant?

(A) Adding $H_2SO_4(g)$
(B) Increasing the concentration of $KMnO_4(aq)$
(C) Removing $CO_2(aq)$
(D) Removing $MnSO_4(aq)$

268. A kinetic study of the following reaction led to a proposed mechanism:

$$Cl_2(g) + H_2O(g) \rightarrow HCl(g) + HClO(g)$$

The proposed mechanism was:

$$Cl_2 \leftrightarrows 2 Cl\bullet \qquad \text{fast}$$
$$Cl\bullet + H_2O \rightarrow HCl + \bullet OH \qquad \text{slow}$$
$$Cl\bullet + \bullet OH \rightarrow HClO \qquad \text{fast}$$

Which of the following observations would support this mechanism?

(A) Learning that a free radical scavenger inhibits the reaction
(B) The disappearance of Cl_2
(C) The determination that the rate law is Rate = $k[Cl_2]$
(D) The determination that a decrease in temperature caused the reaction to slow

269. Referring to the proposed mechanism in question 268, the first step is promoted by light. Which of the following would have the *least* effect on the rate of the reactions in the mechanism?

(A) The addition of a substance that inhibits H–O bond cleavage
(B) A change in the light intensity
(C) The addition of a substance that removes free radicals
(D) The addition of a small amount of argon gas

270. Which of the following correctly describes the two energy profiles shown below?

(A) Profile I is endothermic, and profile II is endothermic.
(B) Profile I is exothermic, and profile II is exothermic.
(C) Profile I is exothermic, and profile II is endothermic.
(D) Profile I is endothermic, and profile II is exothermic.

271. Which of the following energy profiles illustrates a reaction that requires energy to proceed?

(A)

(C)

(B)

(D)

272. Which of the following energy profiles corresponds to a mechanism where the first step is the rate-determining step?

(A)

(C)

(B)

(D)

273. The following energy profile is for an exothermic reaction with $\Delta H = -145$ kJ•mol^{-1} and an activation energy of 285 kJ•mol^{-1}. What is the activation energy of the reverse reaction?

(A) 285 kJ•mol^{-1}
(B) 430 kJ•mol^{-1}
(C) 140 kJ•mol^{-1}
(D) 145 kJ•mol^{-1}

274. For the following reaction

$$2\ H_2O_2(aq) \rightarrow 2\ H_2O(aq) + O_2(g)$$

In the absence of a catalyst, the rate of the reaction is: $-\frac{\Delta[H_2O_2]}{\Delta t} = 1.0 \times 10^{-8}$ M/s. However, in the presence of hemoglobin, which acts as a catalyst, the rate increases from 1.0×10^{-8} M/s to 2.0×10^{-1} M/s. What is the rate of oxygen production for the hemoglobin-catalyzed reaction?

(A) 1.0×10^{-1} M/s
(B) 2.0×10^{-1} M/s
(C) 1.0×10^{-8} M/s
(D) 5.0×10^{-2} M/s

275. Consider the following energy profile:

It illustrates a reaction in which:

(A) The forward reaction is faster than the reverse reaction
(B) The reverse reaction is faster than the forward reaction
(C) The forward and reverse reactions proceed at the same rate
(D) It is impossible to predict the relative rates of the forward and reverse reactions

276. Which of the following will *not* change the rate of the following reaction?

$$CaCO_3(s) + 2\ HCl(g) \rightarrow CaCl_2(aq) + H_2O(l) + CO_2(g)$$

(A) Concentration of $CaCl_2(aq)$
(B) Partial pressure of HCl
(C) Size of $CaCO_3$ particles
(D) Temperature

277. A kinetic study of the following reaction led to a proposed mechanism:

$$Cl_2(g) + H_2O(g) \leftrightarrows HCl(g) + HClO(g)$$

The proposed mechanism was:

$Cl_2 \leftrightarrows 2\ Cl\bullet$	fast
$Cl\bullet + H_2O \rightarrow HCl + \bullet OH$	slow
$Cl\bullet + \bullet OH \rightarrow HClO$	fast

The •OH is an example of:

(A) A promoter
(B) A catalyst
(C) An inhibitor
(D) An intermediate

278. Which feature of the following energy profile would a catalyst change?

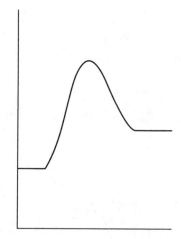

(A) The position at the end of the reaction
(B) The position at the start of the reaction
(C) The height of the peak
(D) Nothing

279. Kinetic studies of a certain reaction led to the following energy profile:

If the reaction were allowed to go to equilibrium, the value of the equilibrium constant would be:

(A) > 1
(B) < 1
(C) = 1
(D) Undetermined

280. In the following energy profile, the activated complex would be at which position?

281. Kinetic studies on biological systems are often done at 37°C because this temperature is the normal body temperature for humans. How would the rate of reaction at 27°C compare to the rate of reaction at 37°C?

(A) One-tenth times that at 37°C
(B) Two times that at 37°C
(C) Ten times that at 37°C
(D) One-half times that at 37°C

282. Which of the following reactions will proceed at the slowest rate?

(A) $A + B \rightarrow C$ $\Delta H = 100$ kJ
(B) $D + E \rightarrow F + G$ $\Delta H = 150$ kJ
(C) $H \rightarrow I + J$ $\Delta H = 250$ kJ
(D) It is impossible to determine.

283. What is the role of Br(g) in the following mechanism?

$Br_2(g) \rightarrow 2\ Br(g)$ slow

$2\ [Br(g) + NO(g) \rightarrow NOBr(g)]$ fast

(A) It is a Lewis base.
(B) It is a catalyst.
(C) It is a product.
(D) It is an intermediate.

284. In a chemical reaction, the rate-determining step is the one that:

(A) Takes the least time
(B) Takes the most time
(C) Is the first step in the mechanism
(D) Is the most complicated step

285. The following table contains data obtained in the study of the reaction A + 2 B + C → X + Y.

Experiment	$[A]_{initial}$	$[B]_{initial}$	$[C]_{initial}$	Initial Rate (M/s)
1	0.1 M	0.1 M	0.1 M	5.0×10^{-2}
2	0.2 M	0.1 M	0.1 M	1.0×10^{-1}
3	0.2 M	0.2 M	0.1 M	1.0×10^{-1}
4	0.1 M	0.1 M	0.3 M	4.5×10^{-1}

What is the rate law for this reaction?

(A) Rate = $k[A][B]^0[C]^2$
(B) Rate = $k[A][B]^2[C]$
(C) Rate = $k[A][B]^0[C]^3$
(D) Rate = $k[A]^2[B][C]$

286. Which of the following factors will increase the value of the rate constant of a reaction?

(A) Removing a catalyst
(B) Decreasing the temperature
(C) Increasing the activation energy
(D) Increasing the temperature

287. For a particular chemical reaction, the rate of reaction is independent of:
    (A) Magnitude of the activation energy
    (B) Overall enthalpy change for the reaction
    (C) Absolute temperature
    (D) Presence or absence of a catalyst

288. The following reaction begins very slowly but rapidly speeds up as Mn^{2+}(aq) forms:

$$3 H_2SO_4(aq) + 2 KMnO_4(aq) + 5 H_2C_2O_4(aq) \rightarrow$$
$$K_2SO_4(aq) + 2 MnSO_4(aq) + 10 CO_2(g) + 8 H_2O(l)$$

One form of the rate of reaction for this overall reaction is:
    (A) Rate $= - (1/2) \Delta[KMnO_4]/\Delta t$
    (B) Rate $= - (2) \Delta[KMnO_4]/\Delta t$
    (C) Rate $= - (1/2) \Delta[MnSO_4]/\Delta t$
    (D) Rate $= - (10) \Delta[CO_2]/\Delta t$

289. A kinetic study of the following reaction led to the determination of the rate law:

$$Cl_2(aq) + H_2O(l) \leftrightarrows HCl(aq) + HClO(aq)$$

Choose the *false* statement:
    (A) There is a change in the oxidation state of chlorine.
    (B) The reaction proceeds more rapidly at a higher temperature.
    (C) Increasing the partial pressure of the chlorine over the solution increases the equilibrium constant.
    (D) Both a weak and a strong acid form in the reaction.

290. A kinetic study of the following reaction led to a proposed mechanism.

$$Cl_2(g) + H_2O(g) \leftrightarrows HCl(g) + HClO(g)$$

A proposed mechanism was:

$Cl_2 \leftrightarrows 2\ Cl\bullet$	fast
$Cl\bullet + H_2O \rightarrow HCl + \bullet OH$	slow
$Cl\bullet + \bullet OH \rightarrow HClO$	fast

Based upon this mechanism, what is the rate law?

(A) Rate = $k[Cl_2]$
(B) Rate = $k[Cl_2]^{1/2}[H_2O]$
(C) Rate = $k[HCl][HClO]$
(D) Rate = $k \frac{[Cl_2][H_2O]}{[HCl][HClO]}$

291. The following reactions were studied under identical conditions:

(1) $Ca(s) + Cl_2(g) \rightarrow CaCl_2(s)$
(2) $Ca(s) + I_2(g) \rightarrow CaI_2(s)$

The rate of reaction for reaction (2) is slower than that for reaction (1). The *best* explanation of this observation is:

(A) The equilibrium constant for reaction (2) is greater than the equilibrium constant for reaction (1).
(B) The activation energy for reaction (2) is greater than the activation energy for reaction (1).
(C) $CaI_2(s)$ is more stable than $CaCl_2(s)$.
(D) $CaCl_2(s)$ is more ionic than $CaI_2(s)$.

292. Nitrogen oxide catalyzes the destruction of ozone in the upper atmosphere. One suggested mechanism is:

$$2 [NO(g) + O_3(g) \rightarrow NO_2(g) + O_2(g)] \qquad\qquad \text{fast}$$
$$2 NO_2(g) \rightarrow N_2O_4(g) \qquad\qquad \text{slow}$$
$$N_2O_4(g) + h\nu \rightarrow 2 NO(g) + O_2(g) \qquad\qquad \text{fast}$$

Which of the following energy profiles applies to this mechanism?

(A)

(C)

(B)

(D)

293. An experiment investigating the oxidation of iodide ion by permanganate ion in acid solution found the rate of disappearance of the permanganate ion to be 2.0×10^{-3} M/s. Determine the rate of appearance of iodine under these conditions:

$$10 I^-(aq) + 2 MnO_4^-(aq) + 16 H^+(aq) \rightarrow 2 Mn^{2+}(aq) + 5 I_2(aq) + 8 H_2O(l)$$

(A) 5.0×10^{-3} M/s
(B) 2.0×10^{-3} M/s
(C) 1.0×10^{-3} M/s
(D) 2.0×10^{-2} M/s

294. Trial studies of a new drug used to treat schizophrenia used an oral dose of 27 mg. The drug remained effective until the level in the bloodstream dropped to about 10% of the initial value. The drug was slowly metabolized in the body, with the metabolism following first-order kinetics with a half-life of 4 hours. How long after the initial administration of the drug would it be necessary to administer a second dose?

(A) 24 hours
(B) 4 hours
(C) 12 hours
(D) 8 hours

CHAPTER 8

Equilibrium

Questions 295–299 refer to the following passage.

Chemical equilibrium is so very important because, in principle, all reactions are reversible. Thus, in addition to reactants forming products, it is possible for products to reform the original reactants. In the first half of this book, the assumption was that reactions proceeded completely from reactants to products. You encountered an exception in Chapter 4 concerning weak electrolytes. In Chapter 7, the concept of chemical equilibrium was described because the reaction has both an activation energy for the forward reaction and an activation energy for the reverse reaction. The difference in these activation energies indicates that the forward and reverse reactions initially proceed at different rates. These rates will change as the concentrations of the reactants and products change. One reaction is initially faster, as it consumes chemical material faster than the slower reaction. The faster reaction slows down as the amounts of its reactants decrease. The rate of the slower reaction increases as the concentration of its reactants increases.

A **chemical equilibrium** is established when the forward reaction is proceeding at the same rate as the reverse reaction; thus, no apparent change in concentration occurs. However, the forward and reverse reactions never stop. Since both of the reactions are still occurring, this is an example of a dynamic equilibrium. At equilibrium, the stability of the system has reached a maximum and there is no net change in the concentrations of the reactants and products.

A double arrow, $\leftrightarrows$, indicates the presence of a chemical equilibrium. For a weak electrolyte, such as nitrous acid (a weak acid), the equilibrium equation is:

$$HNO_2(aq) \leftrightarrows H^+(aq) + NO_2^-(aq)$$

This equation indicates that a nitrous acid solution contains not only hydrogen ions and nitrite ions, but also some nitrous acid molecules.

All equilibriums, whether homogeneous or heterogeneous, are dynamic equilibriums. In a dynamic equilibrium, the reaction is simultaneously proceeding in both directions at equal rates. Thus, there is no net change in the concentrations even though the substances are still reacting. Le Châtelier's principle governs all dynamic equilibriums. **Le Châtelier's principle** states that applying a stress to a system in equilibrium causes the system to respond in order to counter the stress, which leads to the establishment of a new equilibrium. A "stress" is virtually any change imposed upon a system in equilibrium. There are four potential types of stress: concentration changes, volume/pressure changes, temperature changes, and addition of a catalyst.

295. The following equilibrium is important in the industrial preparation of sulfuric acid:

$$2 SO_2(g) + O_2(g) \leftrightarrows 2 SO_3(g)$$

A mixture of sulfur dioxide and oxygen is quickly introduced to a reaction chamber containing a catalyst. After the system comes to equilibrium, the partial pressure of sulfur dioxide is 10.0 atm, the partial pressure of oxygen is 5.0 atm, and the partial pressure of sulfur trioxide is 10.0 atm. What is the value of the equilibrium constant, K_p?

(A) 5.00
(B) 0.20
(C) 10.0
(D) 0.10

296. The following equilibrium is important in the industrial preparation of sulfuric acid:

$$2 SO_2(g) + O_2(g) \leftrightarrows 2 SO_3(g)$$

In one experiment, a mixture of sulfur dioxide and oxygen is quickly introduced to a reaction chamber. After the system comes to equilibrium, the partial pressure of the gases allows the determination of the equilibrium constant. In a second experiment, the same mixture plus a catalyst is quickly introduced to the same reaction chamber. After the system comes to equilibrium, the partial pressure of the gases allows the determination of the equilibrium constant. How do the equilibrium constants determined in the two experiments compare?

(A) They are both the same.
(B) The value for the catalyzed experiment was higher.
(C) The value for the uncatalyzed experiment was higher.
(D) It is impossible to determine from the information given.

297. The following equilibrium is important in the industrial preparation of sulfuric acid:

$$2\ SO_2(g) + O_2(g) \leftrightharpoons 2\ SO_3(g)$$

A two-to-one mixture of sulfur dioxide and oxygen is quickly introduced to a reaction chamber containing a catalyst. The initial pressure of the mixture is 30.0 atm. After the system comes to equilibrium, the total pressure is 25.0 atm. What is the partial pressure of sulfur trioxide at equilibrium?

(A) 5.0 atm
(B) 10.0 atm
(C) 25.0 atm
(D) 7.5 atm

298. An industrial chemist was reexamining the following equilibrium.

$$N_2(g) + 3\ H_2(g) \leftrightharpoons 2\ NH_3(g)$$

She performed two experiments in the course of her investigation. Both experiments began with identical quantities of N_2 and H_2 in identical containers at the same temperature and pressure. However, before beginning the second experiment, a catalyst was added to the system. Both systems go to equilibrium. At the end of each experiment:

(A) The concentrations of all reactants and products are the same in each system
(B) The concentrations of the products are higher in the system containing the catalyst
(C) The concentrations of the reactants are higher in the system without a catalyst
(D) It is not possible to determine the relative amounts of reactants and products

299. The reaction of nitrogen gas with hydrogen, in the presence of a catalyst, will reach an equilibrium with $\Delta H = -17$ kJ/mol NH_3. The equilibrium reaction is:

$$N_2(g) + 3\ H_2(g) \leftrightharpoons 2\ NH_3(g)$$

Increasing the temperature of this system in equilibrium will:

(A) Increase the kinetic energy of the molecules
(B) Increase the concentration of ammonia
(C) Not change anything
(D) Decrease the total pressure

300. The K_{sp} of AgCN is 2.2×10^{-16}, and the K_{sp} of AgOCN is 2.3×10^{-7}. Which of the following statements about these two compounds is true?

(A) The value of ΔG for the dissolution of AgCN is more positive than that of AgOCN.

(B) AgCN is more soluble than AgOCN.

(C) Both compounds are more soluble in a 0.1 M $AgNO_3$ solution than in pure water.

(D) AgOCN is more soluble than AgCN.

301. The K_{sp} of $AgC_2H_3O_2$ is 4×10^{-3}. What is the minimum acetate ion concentration needed to initiate the precipitation of $AgC_2H_3O_2$ from a 1×10^{-3} M Ag^+ solution?

(A) 4 M

(B) 1 M

(C) 4×10^{-3}

(D) 1×10^{-3} M

302. Calcium hydroxide [$Ca(OH)_2$] is slightly soluble in water. Which of the following *best* represents the expression for the solubility product of calcium hydroxide?

(A) $K_{sp} = [CaOH^-][OH^-]$

(B) $K_{sp} = [Ca(OH)_2]$

(C) $K_{sp} = [Ca^{2+}][OH^-]^2$

(D) $K_{sp} = [Ca^{2+}][(OH)^{2-}]$

303. A student has a beaker with about 250 mL of solution and some slightly soluble calcium sulfate, $CaSO_4$, on the bottom. The student then adds 25 mL of a sodium sulfate solution. What happens?

(A) More calcium sulfate dissolves.

(B) The calcium ion concentration increases.

(C) There is a decrease in the pH.

(D) There will be an increase in the amount of undissociated calcium sulfate.

304. What would be the result of increasing the pressure upon the following equilibrium system?

$$CH_4(g) + H_2O(g) \leftrightarrows CO(g) + 3\,H_2(g)$$

(A) The amount of CO would increase.
(B) The amount of H_2O would decrease.
(C) The amount of CH_4 would increase.
(D) There would be no change.

305. Which of the following statements is *false* in the study of equilibrium reactions?

(A) The presence of a catalyst does not change the value of the equilibrium constant.
(B) The addition of a reactant or product to a homogeneous gaseous equilibrium is important.
(C) The addition of a solid reactant to an equilibrium causes no change.
(D) The presence of a catalyst will increase the value of the equilibrium constant.

306. Some kidney stones consist primarily of calcium oxalate. These kidney stones form only in certain patients. These patients:

(A) Have a low blood pH
(B) Have a high blood pH
(C) Also have acidosis
(D) Have a calcium deficiency

307. A 0.250 μg sample of insoluble CoS containing only radioactive cobalt-60 is added to 10 mL of water. After achieving equilibrium, 5 μg of soluble $CoCl_2$ containing nonradioactive cobalt-59 is dissolved in the solution. Which of the following is true once the system reaches equilibrium a second time?

(A) The radioactive decay rate of cobalt-60 increases.
(B) There will be more cobalt-60 in solution.
(C) There will be the same amount of cobalt-60 in solution.
(D) The solid will now be a mixture of CoS containing Co-60 and CoS containing Co-59.

308. The addition of $Ba(NO_3)_2$ to a solution produces a precipitate. The precipitate could be:

(A) Barium sulfate
(B) Potassium nitrate
(C) Zinc nitrate
(D) Barium chloride

309. The symbol "$\leftrightharpoons$" indicates:

(A) The presence of water
(B) A reversible reaction
(C) Equilibrium
(D) A buffer solution

310. The autoionization of water $[2\ H_2O(l) \leftrightharpoons H_3O^+(aq) + OH^-(aq)]$ is an endothermic process. How does the pH of water change when it is heated?

(A) It decreases.
(B) It increases.
(C) It remains 7.
(D) It changes in an unpredictable manner.

311. The following equilibriums are important in solutions of H_2S:

$$H_2S(aq) \leftrightharpoons H^+(aq) + HS^-(aq) \quad K_{a1} = 1.0 \times 10^{-7}$$

$$HS^-(aq) \leftrightharpoons H^+(aq) + S^{2-}(aq) \quad K_{a2} = 1.3 \times 10^{-13}$$

The addition of sulfuric acid to a saturated solution of H_2S will *not* cause the:

(A) pH to become lower
(B) Concentration of undissociated H_2S to increase
(C) Release of hydrogen gas
(D) K_{a2} equilibrium to shift to the left

312. The following equilibrium will form phosgene from carbon monoxide and chlorine:

$$CO(g) + Cl_2(g) \leftrightarrows COCl_2(g)$$

What will happen to the equilibrium if there is a sudden increase in the volume of the system after equilibrium is established?

(A) The equilibrium will shift to the right.
(B) The equilibrium will shift to the left.
(C) The equilibrium will not shift.
(D) It impossible to determine what will happen to the equilibrium.

313. The following exothermic reaction will form phosgene from carbon monoxide and chlorine:

$$CO(g) + Cl_2(g) \leftrightarrows COCl_2(g)$$

An increase in temperature will cause:

(A) An unknown change
(B) The equilibrium to shift to the right
(C) No change in the equilibrium
(D) The equilibrium to shift to the left

314. The content of air exiting the lungs during normal breathing is:

(A) Higher in carbon dioxide and lower in oxygen relative to air
(B) Higher in carbon dioxide and higher in oxygen relative to air
(C) Lower in carbon dioxide and lower in oxygen relative to air
(D) Lower in carbon dioxide and higher in oxygen relative to air

315. A mixture is placed in a cylinder with a moving piston. The mixture achieves equilibrium, and then the piston moves in, reducing the volume. Which of the following equilibriums will undergo the greatest change?

(A) $Ni(s) + 4 CO(g) \leftrightarrows Ni(CO)_4(g)$
(B) $CaCO_3(s) \leftrightarrows CaO(s) + CO_2(g)$
(C) $Fe^{3+}(aq) + 6 CN^-(aq) \leftrightarrows [Fe(CN)_6]^{3-}(aq)$
(D) $2 H_2(g) + O_2(g) \leftrightarrows 2 H_2O(g)$

316. The following reaction is at equilibrium:

$$Cl_2(aq) + H_2O(l) \leftrightarrows HCl(aq) + HClO(aq)$$

A quantity of HCl(g) dissolves in the mixture and triples the concentration of HCl. What effect will the HCl have on the equilibrium?

(A) The reaction quotient will be one-third the equilibrium constant.
(B) The reaction quotient will be triple the equilibrium constant.
(C) The reaction quotient will be equal to the equilibrium constant.
(D) The reaction quotient will be equal to 3.

317. Some chlorine oxides will dissolve in water to produce acids. Two examples are:

$$Cl_2O(g) + H_2O(l) \leftrightarrows 2\ HClO(aq)$$
$$Cl_2O_5(g) + H_2O(l) \leftrightarrows 2\ HClO_3(aq)$$

Why is Cl_2O_5 more soluble than Cl_2O?

(A) $HClO_3$ is a strong acid, and its dissociation shifts its equilibrium to the right.
(B) $HClO_3$ is a strong acid, and its dissociation shifts its equilibrium to the left.
(C) HClO is a strong acid, and its dissociation shifts its equilibrium to the right.
(D) HClO is a strong acid, and its dissociation shifts its equilibrium to the left.

318. The solubility constant of calcium fluoride, CaF_2, is 4.0×10^{-11}. Will calcium fluoride precipitate from a solution prepared by adding 10.0 mL of 0.01 M $CaCl_2$ and 20.0 mL of 0.01 M HF to sufficient water to make 1.0 L of solution?

(A) Yes
(B) No
(C) Yes, if the temperature is increased
(D) There is insufficient data.

319. Which of the following would be most effective at decreasing the solubility of AgCN in water?

(A) 1 M HCN
(B) 1 M NaCN
(C) 1 M $Ca(CN)_2$
(D) 1 M $C(CN)_3COOH$

320. Which of the following would have a minimal effect on the following equilibrium?

$$CaCO_3(s) \leftrightarrows CaO(s) + CO_2(g)$$

(A) Adding CaO to the container
(B) Adding CO_2 to the container
(C) Heating the mixture in the container
(D) Decreasing the volume of the container

321. The following equilibrium occurs at a certain temperature:

$$CH_4(g) + H_2O(g) \leftrightarrows CO(g) + 3\ H_2(g)$$

Which of the following would have a minimal effect on the equilibrium concentrations?

(A) Changing the temperature
(B) Removing CO
(C) Adding CH_4
(D) Adding a catalyst

322. The following equilibrium occurs at a certain temperature:

$$N_2(g) + O_2(g) \leftrightarrows 2\ NO(g)$$

Which of the following would have a minimal effect on the equilibrium concentrations?

(A) Changing the temperature
(B) Changing the pressure
(C) Changing the amount of NO
(D) Changing the amount of N_2

323. The following equilibrium occurs at a certain temperature:

$$N_2(g) + O_2(g) \leftrightarrows 2\ NO(g)$$

What is the value of the equilibrium constant, K_p, of this reaction if the equilibrium concentrations are $[N_2] = 0.24$ atm, $[O_2] = 0.48$ atm, and $[NO] = 0.48$ atm?

(A) 10
(B) 2.0
(C) 0.10
(D) 0.50

324. A solution of $ZnCO_3$ ($K_{sp} = 1.0 \times 10^{-10}$) is at equilibrium. The addition of a small quantity of Na_2CO_3 will:

    (A) Cause more $ZnCO_3$ to dissolve
    (B) Cause more $ZnCO_3$ to precipitate
    (C) Increase the value of the solubility constant
    (D) Decrease the value of the equilibrium constant

325. The solubility of AgCN in water is about 1.0×10^{-8} mol/L. What is the solubility product constant for AgCN?

    (A) 1.0×10^{-24}
    (B) 1.0×10^{-8}
    (C) 1.0×10^{-16}
    (D) 2.0×10^{-8}

326. $CaSO_4$ is added to a 0.1 M Na_2SO_4 solution. The solubility of $CaSO_4$ is:

    (A) Lowered due to the common ion effect
    (B) Increased due to the common ion effect
    (C) Unchanged because Na^+ compounds are always soluble
    (D) Lowered because the Na_2SO_4 combines with most of the water, leaving little to dissolve the $CaSO_4$

327. What is the *best* way of expressing the equilibrium constant, K_c, for the following reaction?

$$2\ NO(g) + O_2(g) \leftrightharpoons 2\ NO_2(g)$$

    (A) $[NO_2]^2 / [NO]^2[O_2]$
    (B) $2[NO_2] / 2[NO][O_2]$
    (C) $[NO]^2[O_2] / [NO_2]^2$
    (D) $[NO_2] / [NO][O_2]$

328. Which of the following would increase the production of $H_2(g)$ in the following equilibrium?

$$CH_4(g) + H_2O(g) \leftrightharpoons CO(g) + 3\ H_2(g)$$

    (A) Adding CO
    (B) Removing H_2
    (C) Removing CH_4
    (D) Increasing the pressure

329. Barium oxalate, BaC_2O_4, has a solubility constant of 1.6×10^{-6}. Barium oxalate will be most soluble in which of the following?

(A) 0.1 M $Ba(NO_3)_2$
(B) Pure water
(C) 0.1 M $NaNO_3$
(D) 0.1 M HNO_3

330. The solubility product constant for cupric arsenate is 10^{-37}. What is the approximate arsenate concentration in a saturated solution of cupric arsenate?

(A) 10^{-37} M
(B) 10^{-8} M
(C) 10^{-7} M
(D) 10^{-18} M

331. A 0.250 g sample of insoluble AgI containing only radioactive iodine-131 is added to 1 L of water. After achieving equilibrium, 5 g of soluble KI containing nonradioactive iodine-127 is dissolved in the solution. Which of the following is true once the system reaches equilibrium a second time?

(A) The radioactive decay rate of iodine-131 increases.
(B) There will be less iodine-131 in solution.
(C) There will be the same amount of iodine-131 in solution.
(D) There will be more iodine-131 in solution.

332. What is the concentration of a solution prepared by adding 300 g of calcium carbonate to 500 mL of water?

(A) 3 M
(B) 6 M
(C) 0.6 M
(D) < 0.6 M

CHAPTER 9

Acids and Bases

Questions 333–340 refer to the passage below.

As the body uses the oxygen carried by the hemoglobin in the blood to oxidize glucose, both carbon dioxide and hydrogen ions, H^+(aq), are produced. If the concentration of hydrogen ions is allowed to increase, the blood pH will drop lower than the normal pH of 7.4. A condition called acidosis results. This can be very harmful, because many of the reactions in the body are pH dependent. Ideally, the pH of the blood should be maintained within a 7.2 to 7.6 range. Values that are much lower or higher than this can result in death. It is the role of the buffers in the blood to maintain this pH range.

The most important buffer system in the body acting to maintain a normal blood pH is the carbonic acid/bicarbonate ion buffer shown below:

$$H_2CO_3(aq) \leftrightarrows HCO_3^-(aq) + H^+(aq)$$

The K_a for this system is approximately 7.9×10^{-7} at normal body temperature. In order for a buffer system to have as much buffering capacity as possible for both an acid and a base, the ratio of acid to base should be 1:1, so that the pH of the buffer = pK_a. In this case, that would give a pH of 6.1, far outside the normal blood pH of 7.4. There must be another factor we must take into consideration.

Recall that carbonic acid is rather unstable in water, decomposing to carbon dioxide and water:

$$H_2CO_3(aq) \leftrightarrows CO_2(g) + H_2O(l)$$

Therefore, a more complete representation of the significant blood buffering system can be expressed as:

$$CO_2(g) + H_2O(l) \leftrightarrows H_2CO_3(aq) \leftrightarrows HCO_3^-(aq) + H^+(aq)$$

Looking at both of these equilibriums, it becomes apparent that the amount of carbon dioxide will affect the buffer system. In fact, an equilibrium expression in terms of the hydrogen ion concentration can be derived relating the carbon dioxide and bicarbonate ion concentration:

$$[H^+] = \left[\frac{K_a}{K}\right]\left[\frac{[CO_2]}{[HCO_3^-]}\right]$$

where K_a is the ionization constant for carbonic acid and K is the equilibrium constant associated with the decomposition of the carbonic acid. This relationship can then be expressed in terms of pH:

$$pH = pK - \log\left[\frac{[CO_2]}{[HCO_3^-]}\right] \text{ where } K = \left[\frac{K_a}{K}\right]$$

The pH is thus dependent on the ratio of the carbon dioxide to bicarbonate.

333. Which of the following can combine with HNO_2 to form a buffer solution?

(A) HNO_3
(B) NaOH
(C) KNO_3
(D) CH_3OH

334. Which of the following compounds can buffer a solution against both acids and bases?

(A) Na_2CO_3
(B) $NaHCO_3$
(C) NH_4Cl
(D) $CaCl_2$

335. In order to maximize the buffer capacity of a solution, it is *best* to:

(A) Maximize the concentrations of the conjugate acid and conjugate base
(B) Make sure the conjugate acid and conjugate base concentrations are the same
(C) Adjust the pH to near 7
(D) Choose a system where the pK_a is the same as the pK_b

336. The addition of sodium cyanate to a cyanic acid solution causes the pH to:

    (A) Increase due to the common ion effect
    (B) Decrease due to the common ion effect
    (C) Remain constant because a buffer forms
    (D) Remain constant because sodium cyanate is a neutral salt

337. An acid with a K_a of 2×10^{-6} is used to make a buffer. What is the approximate pH of the solution if the buffer has equal concentrations of both the acid and its conjugate base?

    (A) Between 4 and 5
    (B) Between 2 and 3
    (C) Between 3 and 4
    (D) Between 5 and 6

338. Which of the following combinations will yield a buffer solution?

    (A) $NaHCO_3$ and Na_2CO_3
    (B) HCl and NaCl
    (C) CaO and $Ca(OH)_2$
    (D) $C_2H_5OH_2{}^+Cl^-$ and C_2H_5OH

339. A solution prepared by mixing equal volumes of 0.50 M nitric acid with 0.50 M ammonia would have:

    (A) The same pH as a 0.25 M ammonium nitrate solution
    (B) The same pH as 0.50 M ammonium nitrate
    (C) A pH above 7
    (D) The same pH as 1.0 M ammonium nitrate

340. A researcher examining an environmental sample thought to contain chromium oxide added some of the solid to water. The solution formed had a pH of 1.0. Which of the following oxides of chromium might be present?

    (A) CrO_2
    (B) CrO
    (C) Cr_2O_3
    (D) CrO_3

341. The hydrogen phosphate ion (HPO_4^{2-}) can serve as an acid. Which of the following will result from the acidic dissociation of the hydrogen phosphate ion?

(A) PO_4^{3-}
(B) PO_4^{2-}
(C) $H_2PO_4^-$
(D) H_3PO_4

342. All the hydrogen halides behave as acids in aqueous solution. Which of the following might be the conjugate base of a hydrogen halide?

(A) At^-
(B) At
(C) OH^-
(D) H^+

343. Which of the following procedures would indicate the presence of carbonate ion in a solution?

(A) The addition of a base to cause the generation of a gas
(B) The addition of an acid to cause the generation of a gas
(C) The addition of NaCl to form a precipitate
(D) The addition of $(NH_4)_2SO_4$ to form a precipitate

344. Which of the following acids is the strongest?

(A) CCl_3COOH
(B) $CHCl_2COOH$
(C) $CH_2ClCOOH$
(D) CH_3COOH

345. The pH of a 10^{-2} M solution of HNO_3 is:

(A) < 2
(B) 2
(C) > 2
(D) Unknown

346. The pH of blood is about 7, and the pH of stomach acid is about 1. The hydrogen ion concentration in stomach acid is how many times greater than that in blood?

(A) 10^6
(B) 10^{-6}
(C) 6
(D) -6

347. Ten milliliters of a dilute hydrochloric acid solution with a pH of 6 is diluted to 1,000 mL. The pH of the diluted solution is:

(A) > 7
(B) < 7
(C) 7
(D) There is not enough information to answer.

348. Water from abandoned lead mines is often acidic. Which of the following substances might cause the water to be acidic?

(A) SO_2
(B) SiO_2
(C) $CaCO_3$
(D) PbS

349. The pH of blood is about 7.4. Which of the following is *not* true concerning blood?

(A) $CaCO_3$ is more soluble in blood than in water.
(B) The hydrogen ion concentration is less than the hydroxide ion concentration.
(C) The pOH is about 6.6.
(D) Organic acids will be more soluble in blood than in pure water.

350. An acid-base indicator may use any of the following functional groups *except*:

(A) Sulfonic acid
(B) Amine
(C) Carboxylic acid
(D) Alcohol

351. The K_b of pyridine, C_5H_5N, is 2×10^{-9}. What is the approximate pH of a 1 M solution of pyridine?

(A) 9
(B) 3
(C) 12
(D) 5

352. Weak acids, such as formic acid (CHOOH), exist in equilibrium in solution. The equilibrium for formic acid is:

$$CHOOH(aq) \leftrightarrows H^+(aq) + CHCOO^-(aq)$$

The addition of potassium hydroxide to a formic acid solution will cause all of the following *except*:

(A) An increase in the ability of the solution to conduct electricity
(B) A shift in the equilibrium to the right
(C) A decrease in the pH
(D) An increase in the concentration of the formate ion

353. The addition of solid potassium hydroxide to a 0.1 M nitric acid solution results in the neutralization of 99% of the acid. What is the pH of the solution?

(A) 7
(B) 1
(C) 3
(D) 14

354. Which of the following is an example of a polyprotic acid?

(A) $PO(OH)_3$
(B) CH_3COOH
(C) NH_3
(D) $NO_2(OH)$

355. The process producing water from H^+ and OH^- is very exothermic. This means that when heating water, the value of K_w:

(A) Will change, but it is necessary to know the pressure to know how
(B) Decreases
(C) Remains constant
(D) Increases

356. A water sample was contaminated with a sodium halide. The pH of the water sample was greater than 7. Which sodium halide is the most likely contaminant?

(A) NaF
(B) NaCl
(C) NaBr
(D) NaI

357. At room temperature, the pH of household ammonia is about 11. The pH of vinegar is about 3. The $[H^+]$ in vinegar is how much higher than that in household ammonia?

(A) 1/8
(B) 8
(C) 10^8
(D) 800

358. The K_a (ionization constant) value for a strong acid is:

(A) $K_a = 1$
(B) $K_a < 1$
(C) $K_a > 1$
(D) K_a depends on the identity of the strong acid.

359. Arsenic acid, H_3AsO_4, is a weak triprotic acid. Which of the following is true concerning the concentrations in a 0.10 M arsenic acid solution?

(A) $[AsO_4^{3-}] > [HAsO_4^{2-}]$
(B) $[H^+] > [H_2AsO_4^-]$
(C) $[H^+] < [H_2AsO_4^-]$
(D) $[AsO_4^{3-}] = [H_3AsO_4]$

360. A solution is prepared by dissolving 1.0 mole of a compound in 500.0 mL of water. Later the solution was diluted to 1.500 L. The pH of the resultant solution was 4.3. What is the relationship between the K_a of the compound and the K_b of its conjugate base?

(A) $K_a < K_b$
(B) $K_a > K_b$
(C) $K_a = K_b$
(D) $K_a = 1/K_b$

361. The following bond energies and K_{a1} values were measured for compounds with the general formula H_2X, where X is an element in the oxygen family:

Bond	Bond Energy (kJ mol⁻¹)	K_{a1}
H–O	428	—
H–S	344	1.0×10^{-7}
H–Se	305	1.3×10^{-4}
H–Te	268	2.3×10^{-3}

What is the relationship between the bond energies in the table, the electronegativity, and the K_a values for the H_2X compounds?

(A) The acidity increases as the bond energy and electronegativity decrease.
(B) The acidity increases as the bond energy decreases and the electronegativity increases.
(C) The acidity increases as the bond energy increases and the electronegativity decreases.
(D) The acidity decreases as the bond energy and electronegativity decrease.

362. Which of the following *best* represents the acidity of the hydrogen halides?

(A) HF = HCl > HBr = HI
(B) HF > HCl > HBr > HI
(C) HI > HBr > HCl > HF
(D) HI > HBr > HCl = HF

363. Given the following K_a values:

Acid	K_a
HClO (hypochlorous acid)	3.5×10^{-8}
$HC_3H_3O_3$ (pyruvic acid)	1.4×10^{-4}

Which of the following would be protonated to the greater extent by the addition of 10.0 mL to a 0.10 M solution of ClO^- or $C_3H_3O_3^-$?

(A) ClO^- because it is a stronger base than $C_3H_3O_3^-$
(B) ClO^- because it is a weaker base than $C_3H_3O_3^-$
(C) $C_3H_3O_3^-$ because it is a stronger base than ClO^-
(D) $C_3H_3O_3^-$ because it is a weaker base than ClO^-

364. When dissolved in water, strong acids appear equal in strength. However, differences in strength are noted in other solvents. Which would be the *best* solvent to differentiate two strong acids?

(A) $NaOH(aq)$
(B) $NH_3(l)$
(C) $(CH_3)_3N(l)$
(D) $HC_2H_3O_2(l)$

365. Amino acids, the constituents of proteins, are molecules with a central carbon atom attached to an amine group, a carboxylic acid group, hydrogen, and an R group. The identity of the R group identifies the amino acid. The following equilibriums are important to all amino acids.

Acid form   Zwitterion form   Base form

Which of the following *best* describes the behavior of amino acids?

(A) They are weak acids.
(B) They are amphoteric.
(C) They are weak bases.
(D) They are ionizable.

366. The reaction of 15.00 mL of 1.5 M $Ba(OH)_2$ with 30.00 mL of 1.5 M HNO_3 in a calorimeter resulted in a 15.0°C increase in the temperature. What would be the pH of the solution after the mixture cooled to the original temperature?

(A) > 7
(B) < 7
(C) 7
(D) Impossible to determine

367. Identify the Brønsted-Lowry base in the following reaction of importance in organic synthesis.

$$HNO_3(aq) + H_2SO_4(aq) \rightarrow (NO_2^+)(HSO_4^-)(aq) + H_2O(l)$$

(A) $HNO_3(aq)$
(B) $H_2SO_4(aq)$
(C) $(NO_2^+)(HSO_4^-)(aq)$
(D) $H_2O(l)$

368. The K_{a1} of ascorbic acid, $H_2C_6H_6O_6$, is 1.0×10^{-6}, and K_{a2} for this acid is 2.0×10^{-12}. What is the approximate concentration of hydrogen ions in a 1 M solution of ascorbic acid?

(A) 1.0×10^{-6} M
(B) 1.0×10^{-3} M
(C) 2.0×10^{-3} M
(D) 1 M

369. What is the pH of a 0.010 M HNO_3 solution?

(A) 1.0
(B) 0.010
(C) 2.0
(D) 0.0

370. Which of the following is *not* a weak base?

(A) CH_3NH_2
(B) NH_3
(C) KOH
(D) F^-

371. Which of the following solutes will produce a basic solution when dissolved in water?

(A) C_2H_5OH
(B) H_3PO_4
(C) NaCl
(D) Na_2CO_3

372. Which of the following statements is *not* true?

(A) HCO_3^- is the conjugate base of H_2CO_3.
(B) CH_3COOH is the conjugate acid of CH_3COO^-.
(C) HNO_3 is the conjugate acid of NO_2^-.
(D) NH_3 is the conjugate base of NH_4^+.

373. The following equilibriums are important in solutions of H_2S:

$$H_2S(aq) \leftrightarrows H^+(aq) + HS^-(aq) \quad K_{a1} = 1 \times 10^{-7}$$

$$HS^-(aq) \leftrightarrows H^+(aq) + S^{2-}(aq) \quad K_{a2} = 1.3 \times 10^{-13}$$

The pK_a of H_2S is about:

(A) 20
(B) 13
(C) 7
(D) 1

374. Which of the following compounds will increase the hydroxide ion concentration when dissolved in water?

(A) NH_4Cl
(B) CH_3CH_2OH
(C) K_3PO_4
(D) $NaCl$

375. Which of the following indicators would be the *best* one to use in an acid-base titration when adding a strong base from a burette to a weak acid?

Indicator	Color Change	pH Range
Thymol blue	yellow-blue	7.4–9.2
Nitrazine yellow	yellow-blue-violet	6.0–7.0
Methyl red	red-yellow	4.2–6.2
Methyl violet	green-blue	1.0–1.5

(A) Nitrazine yellow
(B) Thymol blue
(C) Methyl red
(D) Methyl violet

376. The K_a of the methylammonium ion ($CH_3NH_3^+$) is 2×10^{-12}. What is the approximate pH of a solution with a methylamine concentration equal to the methylammonium ion concentration of the same solution?

(A) 6
(B) 3
(C) 12
(D) 8

377. Methemoglobin forms when the Fe^{2+} in hemoglobin undergoes oxidation to Fe^{3+}. The nitrite ion can cause this oxidation. During nitrite oxidation, the nitrogen in the nitrite ion undergoes reduction from 3+ to 0. What is the oxidation state of iron in hemoglobin?

(A) +3
(B) +2
(C) The same as that of nitrogen in the nitrite ion
(D) Impossible to determine

378. Given the following K_a values:

Acid	K_a
$HC_2H_3O_2$ (acetic acid)	1.7×10^{-5}
HCN (hydrocyanic acid)	4.9×10^{-10}
HClO (hypochlorous acid)	3.5×10^{-8}
$HC_3H_3O_3$ (pyruvic acid)	1.4×10^{-4}

Which of the following is the strongest base?

(A) CN^-
(B) $C_2H_3O_2^-$
(C) ClO^-
(D) $C_3H_3O_3^-$

CHAPTER 10

Thermodynamics

Questions 379–382 refer to the passage below.

Calorimetry is the laboratory procedure scientists use to measure the heat released or absorbed during a process such as a chemical reaction. They use an instrument to measure this heat change—a calorimeter.

There are two general types of calorimeters, constant-pressure and constant-volume. The constant-pressure calorimeter is the simplest; one holds the pressure constant by having the reaction chamber open to the atmosphere. This ensures that the pressure remains constant at atmospheric pressure. We use constant-pressure calorimeters to measure enthalpy changes involving solutions, such as in acid-base neutralization reactions, dilution, and so on. Calorimeters of this type may be easily constructed by stacking two polystyrene (Styrofoam) coffee cups together (to insulate the system from the surroundings), providing some means to stir the solution, and having a thermometer to measure the temperature change. The thermometer is not really part of the system. Heat flows to it and is simply recorded by the thermometer.

Not only do the cups prevent heat from escaping into the surroundings, but Styrofoam itself has a very low heat capacity, so that any heat evolved during a reaction remains in the solution and no appreciable amount of heat is lost in heating the cups. Therefore, the heat evolved or absorbed by the process is contained within the solution and recorded by the thermometer. The heat lost or gained by the process is equal to the heat gained or lost by the solution. Knowing the heat capacity of the solution, its mass, and the temperature change allows us to calculate the heat of reaction (q_{rxn}). You can use this same method to measure the heat capacity of a substance.

Scientists commonly use the constant-volume calorimeter, the bomb calorimeter, to measure heats of combustion, the heat associated with the combustion (burning) of substances. We place the sample to be tested into a cup and place it in a heavy-walled reaction vessel called the bomb. An electrical ignition system is in contact with the sample. We remove the air in the bomb and replace it with oxygen. We then place the bomb in an insulated container filled with a known amount of water. We place a thermometer in the water to measure temperature changes; a stirrer circulates the water around the bomb.

At the appropriate time, we ignite the sample electrically, and it burns in the oxygen gas. The heat given off by the sample is absorbed by the water inside the calorimeter and by the calorimeter itself.

379. An inexpensive substitute for a calorimeter is an insulated foam coffee cup. The major reason why a coffee cup is an effective substitute is because it:

   (A) Is nearly adiabatic
   (B) Remains isothermal
   (C) Is lightweight
   (D) Is unreactive

380. An inexpensive substitute for a calorimeter is an insulated foam coffee cup. A major limitation of using a foam coffee cup for a calorimeter is that it will not work effectively under the following condition(s):

   (A) At very high temperatures
   (B) At very high pressure
   (C) At very low pressure
   (D) All of the above

381. The reaction of 25.00 mL of 1.5 M NaOH with 25.00 mL of 1.5 M HNO_3 in a calorimeter resulted in a 15.0°C increase in the temperature. Repeating the experiment with 25.00 mL of 1.5 M acetic acid replacing the HNO_3 would:

   (A) Result in a smaller temperature change because the enthalpy change is smaller
   (B) Result in a larger temperature change because the enthalpy change is larger
   (C) Result in the same temperature change because all acid-base reactions have the same enthalpy change
   (D) Not be predictable based on the information given

382. The following data came from three experiments using a calorimeter:

Experiment 1	Dissolving 8.0 g of NaOH(s) in 100.0 mL of water resulted in a 15°C temperature increase.
Experiment 2	Mixing 50.00 mL of 4.00 M NaOH with 50.00 mL of 4.00 M HNO_3 resulted in a 52°C temperature increase.
Experiment 3	Dissolving 8.0 g of NaOH(s) in 100.0 mL of 2.00 M HNO_3 resulted in a 65°C temperature increase.

Why are the temperature changes in Experiment 2 and Experiment 3 not the same when both experiments used equal numbers of moles and had the same final volume?

(A) The molarities of the HNO_3 solutions were different.
(B) When NaOH(s) dissolves, the process is endothermic.
(C) When NaOH(s) dissolves, the process is exothermic.
(D) The reaction in Experiment 2 did not go to completion.

383. The following reaction will form nitrogen dioxide, an air pollutant, from nitrogen oxide and oxygen:

$$2 \, NO(g) + O_2(g) \rightarrow 2 \, NO_2(g)$$

During this reaction:

(A) There is a decrease in the entropy of the system
(B) There is an increase in the entropy of the system
(C) The entropy of the system remains constant
(D) The change in entropy is impossible to predict

384. The following reaction will form phosgene from carbon monoxide and chlorine:

$$CO(g) + Cl_2(g) \rightarrow COCl_2(g)$$

Given the following thermodynamic data:

Compound	$\Delta H°_f$ (kJ/mol)	$\Delta G°_f$ (kJ/mol)	S° (J/mol·K)
$CO(g)$	−110	−137	198
$Cl_2(g)$	0	0	223
$COCl_2(g)$	−223	−210	290

Which of the following will most likely change the $\Delta G°$ for the reaction?

(A) Increasing the reaction rate
(B) Adding a catalyst
(C) Lowering the activation energy
(D) Changing the temperature

385. Under a specific set of conditions, a reaction is not spontaneous. What must be true about this reaction under these conditions?

(A) ΔH is positive.
(B) ΔS is negative.
(C) ΔG is positive.
(D) $\Delta G = 0$.

386. Which of the following statements is *not* true?

(A) The entropy of gases is usually greater than that of a solid.
(B) Cooling any substance to absolute zero reduces the entropy of the substance to zero.
(C) Under standard conditions, $\Delta H_f°$ *and* $\Delta G_f°$ for any element in its standard state are zero.
(D) STP and standard conditions are not the same.

387. The standard Gibbs free energy of formation for HF(g) is −270 kJ•mol^{-1}. Based upon this information, what is the ΔG_{rxn} for the following reaction under standard conditions?

$$H_2(g) + F_2(g) \rightarrow 2\ HF(g)$$

(A) −540 kJ
(B) −270 kJ
(C) +540 kJ
(D) More information is necessary.

388. LP gas is liquid propane. The combustion of LP gas is important in heating gas grills and homes because this combustion is very exothermic. This means that K_p for the combustion of LP is:

(A) Negative
(B) Positive and small
(C) Positive and large
(D) Zero

389. A sample of hydrogen gas is in an adiabatic system. This means that:

(A) Hydrogen will not react with the sides of the container
(B) The pressure remains constant
(C) The temperature remains constant
(D) There is no heat exchange between the system and the surroundings

390. The following reaction occurred in a system at constant temperature:

$$CaO(s) + CO_2(g) \rightarrow CaCO_3(s)$$

How does the entropy of the system change as the reaction progresses?

(A). It depends on the temperature.
(B) It increases.
(C) It remains the same.
(D) It decreases.

391. The following reaction is at equilibrium:

$$SnCl_2(s) + Cl_2(g) \leftrightarrows SnCl_4(l)$$

A chemist then adds an additional 0.50 mole of $SnCl_2$. In which direction will the reaction proceed?

(A) To the left
(B) To the right
(C) No change
(D) Impossible to determine from the information given

392. It is possible to produce elemental fluorine by the electrolysis of liquid hydrogen fluoride. Hydrogen is the other product. Which of the following observations is true?

    (A) The formation of fluorine-fluorine bonds releases energy, and the reaction is spontaneous.
    (B) The formation of fluorine-fluorine bonds releases energy, and the reaction is nonspontaneous.
    (C) The breaking of hydrogen-hydrogen bonds requires energy, and the reaction is spontaneous.
    (D) The breaking of hydrogen-hydrogen bonds requires energy, and the reaction is nonspontaneous.

393. Which of the following combinations indicates a system with a process that requires cooling to become spontaneous?

    (A) ΔH is negative, and ΔS is positive.
    (B) ΔH is positive, and ΔS is positive.
    (C) ΔH is positive, and ΔS is negative.
    (D) ΔH is negative, and ΔS is negative.

394. What is the change in the internal energy when a sample of helium gas expands? The gas did 1,475 J of work on the surroundings, and the gas cooled through removal of 375 J of heat energy.

    (A) 1,850 J
    (B) −1,100 J
    (C) −1,850 J
    (D) 1,100 J

395. When a liquid solidifies, there is a negative entropy change in the system. The second law of thermodynamics states that for any spontaneous process, the entropy change must be positive. How can a liquid freeze spontaneously when the entropy change of the system is negative?

    (A) The second law of thermodynamics does not apply to phase changes.
    (B) The removal of heat during freezing causes a greater negative enthalpy change in the surroundings.
    (C) There are more microstates available in the solid; therefore, the solid allows more options.
    (D) The first law of thermodynamic supersedes the second law because freezing also involves the transfer of energy.

396. Liquid ammonia is an important industrial chemical. In the isolation of the liquid, ammonia gas condenses in a cold system. Indicate whether the expected signs of ΔS and ΔH are to be positive or negative.

(A) ΔH is negative, and ΔS is negative.
(B) ΔH is positive, and ΔS is positive.
(C) ΔH is negative, and ΔS is positive.
(D) ΔH is positive, and ΔS is negative.

397. The enthalpy change for the vaporization of CCl_4 is higher than the enthalpy change for the fusion of solid CCl_4. Why is this true?

(A) It requires more energy to completely overcome the dipole-dipole forces to convert a liquid to a gas than to partially overcome the forces during fusion.
(B) More energy is required to break the carbon-chlorine covalent bonds during vaporization, and fusion requires less energy to overcome the London dispersion forces.
(C) It requires more energy to completely overcome the London dispersion forces to convert a liquid to a gas than to partially overcome the forces during fusion.
(D) More energy is required to break the carbon-chlorine ionic bonds during vaporization than to overcome the London dispersion forces during fusion.

398. The Gibbs free energy change for a process is positive. What can be said about the equilibrium constant for the process?

(A) K > 1
(B) K < 1
(C) K = 1
(D) Nothing, as the two are unrelated

399. Calculate the work involved in the following reaction when 4.0 g of calcium reacts with excess water to generate hydrogen gas at 0.0°C and 1.00 atm (1 J = 100 L•atm).

$$Ca(s) + 2\ H_2O(l) \rightarrow Ca(OH)_2(aq) + H_2(g)$$

(A) −220 J
(B) +220 J
(C) +100 J
(D) −100 J

400. The following is the energy profile for a reaction:

The overall enthalpy for the reaction is:
(A) I
(B) II
(C) I + II
(D) II – I

401. The following reaction is nonspontaneous and endothermic. What change can make the reaction spontaneous?

$$COCl_2(g) \rightarrow CO(g) + Cl_2(g)$$

(A) It is not possible to make the process spontaneous.
(B) Decrease the volume.
(C) Decrease the temperature.
(D) Increase the temperature.

402. To guarantee that a reaction has a positive free energy change, the enthalpy change must be:
(A) Negative and the entropy change must be negative
(B) Positive and the entropy change must be negative
(C) Positive and the entropy change must be positive
(D) Negative and the entropy change must be positive

403. Due to the greenhouse effect, a sample of gas (the system) absorbs more energy than it emits. This means that q is:
(A) Positive
(B) Negative
(C) Equal to ΔE + work
(D) Both B and C

404. Which of the following reactions will probably have the greatest increase in ΔS_{rxn}?

(A) $C_6H_{12}O_6(s) + 6\ O_2(g) \rightarrow 6\ CO_2(g) + 6\ H_2O(g)$
(B) $H_2O(l) \rightarrow H_2O(g)$
(C) $6\ CO_2(g) + 6\ H_2O(g) \rightarrow C_6H_{12}O_6(s) + 6\ O_2(g)$
(D) $CH_4(g) + 2\ O_2(g) \rightarrow CO_2(g) + 2\ H_2O(g)$

405. Heating a 150.0 g sample of water from 25°C to 75°C results in an increase in the entropy of the water. The main cause for this increase in entropy is that:

(A) The average kinetic energy of the water molecules increases
(B) Most of the hydrogen bonds break
(C) The water vaporizes
(D) The viscosity of the water increases

406. The following high-temperature reaction is useful in the production of elemental barium:

$$4\ BaO(s) + Si(s) \rightarrow Ba_2SiO_4(s) + 2\ Ba(g)$$

Which of the following statements is true about this reaction?

(A) Silicon undergoes oxidation.
(B) The reaction is spontaneous at room temperature.
(C) There is a decrease in entropy.
(D) Ionic BaO changes to molecular Ba_2SiO_4.

407. The following equation represents the reaction of methane gas with a limited amount of oxygen gas:

$$2\ CH_4(g) + 3\ O_2(g) \rightarrow 2\ CO(g) + 4\ H_2O(g)$$

What are the signs of ΔH and ΔS for this reaction?

(A) ΔH is positive, and ΔS is negative.
(B) ΔH is negative, and ΔS is negative.
(C) ΔH is positive, and ΔS is positive.
(D) ΔH is negative, and ΔS is positive.

408. A student builds the apparatus shown here:

For this system, $\Delta E = q + w$, where q is heat and w is work. (Heat lost by the system is negative, while heat gained by the system is positive.) In an experiment, the student began with a 0.50 mole sample of argon gas at 0°C and 1.00 atm. The sample was heated at constant pressure to 273°C, the final volume determined, and the piston locked in place. Next, the system was wrapped with insulation to prevent any heat exchange with the surroundings. Finally, the piston was unlocked. Which of the following is true?

(A) $q > 0$
(B) $q = 0$
(C) $q < 0$
(D) $\Delta T = 0$

409. The following reaction is at equilibrium:

$$SnCl_2(s) + Cl_2(g) \leftrightarrows SnCl_4(l)$$

The standard Gibbs energy of formation for the materials in the reaction are:

$SnCl_2(s)$	−342 kJ/mol
$Cl_2(g)$	0.0 kJ/mol
$SnCl_4(l)$	−440 kJ/mol

What is the value of ΔG?

(A) 0 kJ/mol
(B) −98 kJ/mol
(C) +98 kJ/mol
(D) Impossible to determine from the information given

410. A chemist studies the following reaction:

$$SnCl_2(s) + Cl_2(g) \leftrightarrows SnCl_4(l)$$

The standard enthalpies of formation for the materials in the reaction are:

$SnCl_2(s)$	−325 kJ/mol
$Cl_2(g)$	0.0 kJ/mol
$SnCl_4(l)$	−511 kJ/mol

What is the value of ΔH for the reaction?

(A) +186 kJ
(B) −186 kJ
(C) 0 kJ
(D) −836 kJ

411. For the melting of aluminum, $Al(s) \leftrightarrows Al(l)$, the heat of fusion is 10.0 kJ/mol and the entropy of fusion is 9.50 J/mol•K. Estimate the melting point of Al.

(A) 1,000 K
(B) 730 K
(C) 1,270 K
(D) There is insufficient information.

CHAPTER 11

Electrochemistry

Questions 412–420 refer to the following passage.

The minimum energy necessary to operate an electrolytic cell depends upon the cell potential. The cell potential may be a standard value from the table of standard reduction potentials, or a nonstandard value from the Nernst equation. Unlike that of a galvanic cell, the potentials for electrolytic cells are negative. The anode for an electrolytic cell is still the place where oxidation occurs, but it is positive instead of negative as in a galvanic cell. The cathode of an electrolytic cell remains the site of reduction; however, the cathode is the negative electrode in an electrolytic cell.

It is possible to use electrolytic cells to isolate many elements from their compounds. For some elements, there are cheaper ways to isolate the elements, but for the most active elements, electrolysis is the method of choice. The commercial production of magnesium is an important application of electrolysis.

From the activity series, we know that magnesium is an active metal. Most magnesium compounds are very stable, and isolation of the element by chemical means is energetically impractical. The preparation of magnesium begins with the addition of calcium oxide, CaO, to seawater. Calcium oxide is a typical basic oxide and reacts with water to produce the strong base calcium hydroxide, $Ca(OH)_2$, which reacts with the magnesium ions present in seawater to precipitate magnesium hydroxide, $Mg(OH)_2$. After separating the precipitate from the water, the addition of hydrochloric acid, HCl, to the solid converts the magnesium hydroxide to magnesium chloride, $MgCl_2$. After drying the magnesium chloride, the solid is heated in an electrolytic cell until it melts. The cell operates by the passage of an electric current through the molten magnesium chloride to yield magnesium metal and chlorine gas.

The cathode, reduction half-reaction is:

$$Mg^{2+} + 2\ e^- \rightarrow Mg$$

The anode, oxidation, half-reaction is:

$$2\ Cl^- \rightarrow Cl_2 + 2\ e^-$$

The overall cell reaction is the sum of these half-reactions, or:

$$MgCl_2 \rightarrow Mg + Cl_2$$

If we used an aqueous solution of magnesium chloride in place of the molten magnesium chloride, the products would be hydrogen gas, H_2, chlorine gas, Cl_2, and solid magnesium hydroxide, $Mg(OH)_2$. Why does water make a difference? The presence of water means that we must consider half-reactions in addition to the reduction of magnesium ions and the oxidation of chloride ions. In all electrolytic cells, we must consider all possible half-reactions. In the case of molten magnesium chloride, there are only the two half-reactions seen earlier. If water is present, as when we attempt to electrolyze aqueous solutions, we must also consider the oxidation and reduction of water. The presence of additional substances increases the possible number of half-reactions; we must reduce all of the possible half-reactions to only one oxidation and one reduction.

412. Which of the following statements is true concerning the reaction in an electrolytic cell?

(A) The reaction is spontaneous, and oxidation occurs at the cathode.
(B) The reaction is nonspontaneous, and oxidation occurs at the cathode.
(C) The reaction is spontaneous, and oxidation occurs at the anode.
(D) The reaction is nonspontaneous, and oxidation occurs at the anode.

413. An electrolytic cell contains molten $MgCl_2(l)$. Which of the following statements is true?

(A) Mg^{2+} is reduced at the anode, and Cl^- is oxidized at the cathode.
(B) Mg^{2+} is reduced at the cathode, and Cl^- is oxidized at the anode.
(C) Mg^{2+} is oxidized at the cathode, and Cl^- is reduced at the anode.
(D) Mg^{2+} is oxidized at the anode, and Cl^- is reduced at the cathode.

414. An electrolytic cell, with two inert electrodes, is used to electrolyze an aqueous potassium fluoride solution. The standard reduction potentials are:

$$F_2(g) + 2\ e^- \rightarrow 2\ F^-(aq) \qquad\qquad E° = +2.87\ V$$
$$O_2(g) + 4\ H^+(aq) + 4\ e^- \rightarrow 4\ H_2O(l) \qquad E° = +1.23\ V$$
$$2\ H_2O(l) + 2\ e^- \rightarrow H_2(g) + 2\ OH^-(aq) \qquad E° = -0.83\ V$$
$$K^+(aq) + 1\ e^- \rightarrow K(s) \qquad\qquad E° = -2.93\ V$$

What substance forms at the cathode?

(A) $K(s)$
(B) $H_2(g)$
(C) $O_2(g)$
(D) $F_2(g)$

415. Electrolytic cells are similar to voltaic cells in construction except that electrolytic cells require an external source of electrical energy. Electrolysis is capable of breaking apart many compounds, such as water. However, to break down a nonelectrolyte, such as water, there must also be an inert electrolyte present. In an experiment, water is placed in an electrolytic cell and a few drops of phosphoric acid are added. When the cell is connected to an external power source, gases begin to form at each electrode. What gas forms at the anode, and what gas forms at the cathode?

(A) $O^{2-}(g)$ at the anode and $H^+(g)$ at the cathode
(B) $H_2(g)$ at the anode and $O_2(g)$ at the cathode
(C) $O_2(g)$ at the anode and $PH_3(g)$ at the cathode
(D) $O_2(g)$ at the anode and $H_2(g)$ at the cathode

416. The major source of magnesium metal is the electrolysis of molten magnesium chloride. In aqueous solution, the important half-reactions would be:

$$Mg^{2+}(aq) + 2\ e^- \rightarrow Mg \qquad\qquad E° = -2.37\ V$$
$$Cl_2(g) + 2\ e^- \rightarrow 2\ Cl^-(aq) \qquad\qquad E° = +1.36\ V$$

Assuming the reduction potential in molten magnesium chloride is the same as in aqueous magnesium chloride, what is the minimum voltage necessary for the electrolysis to occur?

(A) $-3.73\ V$
(B) $+3.73\ V$
(C) $-1.01\ V$
(D) $+1.01\ V$

417. How many moles of chlorine gas form during the formation of 2.7 g of aluminum metal by the electrolysis of aluminum chloride?

    (A) 0.15 mol
    (B) 0.10 mol
    (C) 0.30 mol
    (D) 0.050 mol

418. The major method for producing sodium metal is through the electrolysis of molten sodium chloride. Copper, on the other hand, can be isolated by the electrolysis of aqueous copper chloride solutions. Why not produce sodium through the electrolysis of aqueous sodium chloride?

    (A) The reaction is too slow to be viable.
    (B) The process is too expensive.
    (C) The other product, chlorine gas, reacts explosively with sodium metal.
    (D) Sodium metal reacts with water.

419. The electrolysis of molten $InCl_3$ produced 1.5 g In in 100 minutes. The reaction was:

$$2\ InCl_3(l) \rightarrow 2\ In(l) + 3\ Cl_2(g)$$

What was the approximate current used? (1 F = 96,500 C/mol)

    (A) 0.20 A
    (B) 0.50 A
    (C) 2.0 A
    (D) 5.0 A

420. It is possible to isolate fluorine gas through the electrolysis of molten potassium fluoride. In one experiment, an electrolytic cell electrolyzed a sample of molten potassium fluoride. The cell operated at 5.0 V with a current of 2.0 A. How many grams of fluorine gas formed after 17 minutes? (1 F = 96,500 C/mol)

    (A) 0.41 g
    (B) 0.20 g
    (C) 4.1 g
    (D) 2.0 g

421. The compound MgI_2 is most likely:

    (A) An oxidizing agent
    (B) A reducing agent
    (C) Insoluble in water
    (D) A nonelectrolyte

422. The following reaction has a standard cell potential of +1.66 V:

$$2\ Al(s) + 6\ H^+(aq) \rightarrow 2\ Al^{3+}(aq) + 3\ H_2(g)$$

The value of +1.66 V means:

(A) Decreasing the amount of Al(s) will increase the cell potential
(B) The reverse reaction is spontaneous
(C) Increasing the amount of Al(s) will increase the cell potential
(D) The reverse reaction is nonspontaneous

423. Which of the following will show a variation of cell potential with pressure?

   I. $Mg(s) + 2\ H^+(aq) \rightarrow Mg^{2+}(aq) + H_2(g)$
  II. $Mg(s) + Cu^{2+}(aq) \rightarrow Mg^{2+}(aq) + Cu(s)$
 III. $H_2(g) + 2\ Ag^+(aq) \rightarrow 2\ H^+(aq) + 2\ Ag(s)$
 IV. $14H^+(aq) + 6\ Fe^{2+}(aq) + Cr_2O_7^{2-}(aq) \rightarrow 6\ Fe^{3+}(aq) + 2\ Cr^{3+}(aq) + 7\ H_2O(l)$

(A) I and III
(B) II and IV
(C) I and II
(D) III and IV

424. According to the Nernst equation, what will happen to the cell potential of the following reaction after doubling the quantity of solid silver?

$$H_2(g) + 2\ Ag^+(aq) \rightarrow 2\ H^+(aq) + 2\ Ag(s)$$

(A) There will be no change.
(B) It will be doubled.
(C) It will be halved.
(D) It will be decreased by an undetermined amount.

425. Consider the following half-reaction potentials:

$$IO_4^-(aq) + 2\,H^+(aq) + 2\,e^- \rightarrow IO_3^-(aq) + H_2O(l) \qquad E^\circ = +1.65\ V$$

$$2\,SO_3^{2-}(aq) + 3\,H_2O(l) + 4\,e^- \rightarrow \qquad\qquad\qquad E^\circ = -0.58\ V$$
$$\qquad S_2O_3^{2-}(aq) + 6\,OH^-(aq)$$

$$Zn^{2+}(aq) + 2\,e^- \rightarrow Zn(s) \qquad\qquad\qquad\qquad E^\circ = -0.76\ V$$

$$2\,Hg^{2+}(aq) + 2\,e^- \rightarrow Hg_2^{2+}(aq) \qquad\qquad\qquad E^\circ = +0.91\ V$$

Which of the following is the strongest reducing agent?

(A) $Zn(s)$
(B) $IO_4^-(aq)$
(C) $Hg_2^{2+}(aq)$
(D) $SO_3^{2-}(aq)$

426. Consider the following reaction:

$$Ca(s) + Cl_2(g) \leftrightarrows CaCl_2$$

Which of the following is true concerning this reaction?

(A) Ca is an oxidizing agent.
(B) Cl_2 is a reducing agent.
(C) Cl_2 is an oxidizing agent.
(D) $CaCl_2$ is an oxidizing agent.

427. The following apparatus is used to electroplate silver onto a metal surface:

Object to   Ag
be plated

The object to be plated is the:
(A) Cathode, and it is negatively charged
(B) Cathode, and it is positively charged
(C) Anode, and it is negatively charged
(D) Anode, and it is positively charged

428. The standard reduction potential for chlorine involves the following reaction:

$$Cl_2(g) + 2 e^- \rightarrow 2 Cl^-(aq) \quad E° = +1.36 \text{ V}$$

How does this compare to the other halogens?

(A) $F_2 > Cl_2 = Br_2 > I_2$
(B) $F_2 = Cl_2 > Br_2 > I_2$
(C) $F_2 > Cl_2 > Br_2 > I_2$
(D) $F_2 < Cl_2 < Br_2 < I_2$

429. The standard cell potential for the following reaction is $E° = +0.80$ V:

$$H_2(g) + 2 Ag^+(aq) \rightarrow 2 H^+(aq) + 2 Ag(s)$$

What will happen to the cell potential if there is an increase in the pH?

(A) It will decrease.
(B) It will increase.
(C) It will remain the same.
(D) It is not possible to determine.

430. The cathode compartment of a voltaic cell involves the reduction of aluminum ions. The reduction half-reaction for aluminum ions is:

$$Al^{3+} + 3 e^- \rightarrow Al(s) \quad E° = -1.66 \text{ V}$$

How many faradays of electrons must pass through the cell to deposit 9.0 g of aluminum metal?

(A) 1.0
(B) 0.33
(C) 3.0
(D) 9.0

431. When the following redox equation is balanced,

$$H^+(aq) + Cr_2O_7^{2-}(aq) + Fe^{2+}(aq) \rightarrow Cr^{3+}(aq) + Fe^{3+}(aq) + H_2O(l)$$

the coefficient for $Fe^{2+}(aq)$ is:

(A) 9
(B) 6
(C) 3
(D) 1

432. Which of the following compounds can serve only as an oxidizing agent?

(A) K_2CrO_4
(B) CrF_3
(C) CrO_2
(D) CrO

433. Which of the following oxidizing agents does *not* contain hexavalent chromium?

(A) K_2CrO_4
(B) CrO_3
(C) $K_2Cr_2O_7$
(D) $Cr_6(Si_6O_{18})$

434. The following reaction is pH dependent:

$$Cr_2O_7^{2-}(aq) + 2\ OH^-(aq) \leftrightarrows 2\ CrO_4^{2-}(aq) + H_2O$$

Which of the following statements is true?

(A) This is an oxidation.
(B) This is neither an oxidation nor a reduction.
(C) This is a reduction.
(D) This is a disproportionation.

435. The following reaction is spontaneous under standard conditions:

$$8\ H^+(aq) + 2\ MnO_4^-(aq) + 5\ C_2O_4^{2-}(aq) \rightarrow 2\ Mn^{2+}(aq)$$
$$+ 10\ CO_2(g) + 8\ H_2O(l)$$

What can be said about the value of the Gibbs free energy change?

(A) $\Delta G = 0$
(B) $\Delta G > 0$
(C) $\Delta G < 0$
(D) ΔG is undefined.

436. Identify the oxidizing and reducing agents in the following reaction:

$$4\ Cl_2(g) + 4\ H_2O(l) \rightarrow 8\ H^+(aq) + ClO_4^-(aq) + 7\ Cl^-(aq)$$

(A) Oxidizing agent = Cl_2 and reducing agent = H_2O
(B) Oxidizing agent = H_2O and reducing agent = Cl_2
(C) Oxidizing agent = Cl_2 and reducing agent = Cl_2
(D) Oxidizing agent = Cl^- and reducing agent = ClO_4^-

437. A galvanic cell uses the following cell reaction:

$$3 \text{ In}^{3+}(aq) + 2 \text{ Al}(s) \rightarrow 3 \text{ In}^{+}(aq) + 2 \text{ Al}^{3+}(aq)$$

When the cell is operated, the measured standard potential is +1.3 V. The standard reduction potential for $Al^{3+}(aq)$ is –1.7 V. What is the standard reduction potential for the following half-reaction?

$$\text{In}^{3+}(aq) + 2e^- \rightarrow \text{In}^{+}(aq)$$

(A) –0.4 V
(B) +0.4 V
(C) 3.0 V
(D) 0.0 V

438. The reference standard for half-reactions is the standard hydrogen electrode (SHE). The half-reaction for the SHE is:

$$2 \text{ H}^{+}(aq) + 2 \text{ e}^- \rightarrow H_2(aq)$$

Construction of a cell involves 1 M $H^{+}(aq)$, 1 atm $H_2(g)$, and a platinum electrode. The reason the platinum is necessary is that:

(A) It is capable of undergoing reduction
(B) It is resistant to acid
(C) An electrical conductor is necessary
(D) It is easily oxidized

439. Which of the following reactions involves the oxidation of oxygen?

(A) O_2^{2-} to O_2
(B) O_2 to H_2O
(C) OH^- to H_2O
(D) O_2^- to O_2^{2-}

440. In general, any molecule containing 2 very electronegative atoms covalently bonded together will be a good oxidizing agent. Based on this generality, which of the following should be a good oxidizing agent?

(A) CF_4
(B) SO_2
(C) H_2O_2
(D) Na_2O

441. In the following redox reaction, what is the value of n?

$$14 \ H^+(aq) + 6 \ Fe^{2+}(aq) + Cr_2O_7^{2-}(aq) \rightarrow$$
$$6 \ Fe^{3+}(aq) + 2 \ Cr^{3+}(aq) + 7 \ H_2O(l)$$

(A) 1/6
(B) 6
(C) 14
(D) 1

442. Two of the half-reactions for the halogens are:

$$F_2(g) + 2 \ e^- \rightarrow 2 \ F^-(aq) \quad E° = +2.87 \ V$$

$$Cl_2(g) + 2 \ e^- \rightarrow 2 \ Cl^-(aq) \quad E° = +1.36 \ V$$

Based on these data, which of the following statements is true?
(A) Chlorine will oxidize fluorine.
(B) Both halogens are good reducing agents.
(C) Both halogens are good oxidizing agents.
(D) It is easier to oxidize fluorine than chlorine.

443. In a reduction half-reaction:
(A) An element gains 1 or more electrons and the oxidation state of the element increases
(B) An element loses 1 or more electrons and the oxidation state of the element increases
(C) An element gains 1 or more electrons and the oxidation state of the element decreases
(D) An element loses 1 or more electrons and the oxidation state of the element decreases

444. A voltaic cell has the following two half-reactions

$$(1) \ Sn^{2+}(aq) \rightarrow Sn^{4+}(aq) + 2 \ e^-$$

$$(2) \ Fe^{3+}(aq) + e^- \rightarrow Fe^{2+}(aq)$$

Which of the following statements is correct?
(A) Reaction 2 is a reduction, and it occurs at the cathode.
(B) Reaction 1 is a reduction, and it occurs at the cathode.
(C) Reaction 2 is an oxidation, and it occurs at the anode.
(D) Reaction 1 is a reduction, and it occurs at the anode.

445. Given the following half-reactions:

$$Sn^{4+}(aq) + 2 e^- \rightarrow Sn^{2+}(aq) \quad E° = +0.15 \text{ V}$$

$$Hg_2^{2+}(aq) + 2 e^- \rightarrow 2 Hg(l) \quad E° = +0.79 \text{ V}$$

Which of the following is true about the following reaction?

$$2 Hg(l) + Sn^{4+}(aq) \rightarrow Hg_2^{2+}(aq) + Sn^{2+}(aq)$$

(A) $E° = +0.64$ V, and the reaction is spontaneous.
(B) $E° = -0.94$ V, and the reaction is nonspontaneous.
(C) $E° = -0.64$ V, and the reaction is nonspontaneous.
(D) $E° = +0.94$ V, and the reaction is spontaneous.

446. Identify the oxidizing agent in the following reaction:

$$14 H^+(aq) + 6 Fe^{2+}(aq) + Cr_2O_7^{2-}(aq) \rightarrow 6 Fe^{3+}(aq)$$
$$+ 2 Cr^{3+}(aq) + 7 H_2O(l)$$

(A) $Fe^{2+}(aq)$
(B) $Cr_2O_7^{2-}(aq)$
(C) $Fe^{3+}(aq)$
(D) $Cr^{3+}(aq)$

447. What is the standard electrode potential for a galvanic cell using the following half-reactions?

$$IO_4^-(aq) + 2 H^+(aq) + 2 e^- \rightarrow IO_3^-(aq) + H_2O(l) \qquad E° = +1.65 \text{ V}$$

$$2 SO_3^{2-}(aq) + 3 H_2O(l) + 4 e^- \rightarrow S_2O_3^{2-}(aq) + 6 OH^-(aq) \quad E° = -0.58 \text{ V}$$

(A) +1.07 V
(B) +1.65 V
(C) +2.23 V
(D) +3.88 V

448. If it were necessary to use 3 moles of HClO to oxidize 1 mole of Sn^{2+} in a sample to Sn^{4+}, how many moles of ClO_3^- would be necessary to perform the same oxidation?

$$HClO(aq) + H^+(aq) + 2\ e^- \rightarrow Cl^-(aq) + H_2O \quad +1.49\ V$$

$$ClO_3^-(aq) + 6\ H^+(aq) + 6\ e^- \rightarrow Cl^-(aq) + 3\ H_2O \quad +1.45\ V$$

(A) 1
(B) 3
(C) 2
(D) 6

449. An analytical chemist constructs a galvanic cell. The cell uses the following two half-reactions:

$$HNO_2(aq) + H^+(aq) + e^- \rightarrow NO(g) + H_2O(l) \quad E° = +1.00\ V$$

$$Ce^{4+}(aq) + e^- \rightarrow Ce^{3+}(aq) \qquad\qquad\qquad E° = +1.61\ V$$

She then measures the cell voltage under a variety of conditions. Which of the following conditions will yield the greatest voltage?

(A) Standard
(B) Low pH
(C) High pH
(D) Equilibrium

450. How many grams of chromium could be produced by the electrolysis of a $Cr(NO_3)_3$ solution using a current of 2.00 amps for 3.00 hrs?

(A) 3.90 g
(B) 1.29 g
(C) 11.6 g
(D) 7.20 g

CHAPTER 12

Final Review

451. Water-soluble barium compounds are chronic toxins that are retained in the human body for years. Barium retention is due to barium ions mimicking what ion in humans?

(A) Ra^{2+}

(B) K^+

(C) Ca^{2+}

(D) H^+

452. What is the molar mass of $CaSO_4 \cdot 2H_2O$?

(A) 170 g/mol

(B) 140 g/mol

(C) 4,900 g/mol

(D) 160 g/mol

453. Which of the following pairs are compounds in which the halogen atom is in the same oxidation state?

(A) NH_4BrO_4 and $HBrO_3$

(B) Cl_2O and $HClO_2$

(C) Cl_2O_5 and $HClO_3$

(D) I_2O_3 and HIO_4

454. What is the concentration of nitrate ions in a solution prepared by adding 100 mL of 0.50 M aluminum nitrate solution to 0.900 L of water?

(A) 0.080 M

(B) 0.25 M

(C) 0.050 M

(D) 0.15 M

455. Which of the following would have the highest heat of vaporization?

(A) $CH_3CH_2CH_2OH$
(B) $HOCH_2CH_2OH$
(C) $CH_3CH_2CH_2CH_3$
(D) $CH_3CH_2CH_2F$

456. The heat of vaporization does *not* depend upon which of the following?

(A) The external pressure
(B) The strength of the intermolecular forces
(C) The mass of the molecules
(D) The presence of hydrogen bonding

457. Which of the following has the highest melting point?

(A) H_2O
(B) CaO
(C) CH_3OH
(D) $NaCl$

458. Which of the following produces the most moles of ions in aqueous solution?

(A) 1.0 mol $CHCl_3$
(B) 0.30 mol $CaCl_2$
(C) 0.45 mol KCl
(D) 0.25 mol $CeCl_3$

459. Dehydrated patients require intravenous saline solution. Why use saline solution instead of pure water?

(A) The osmotic pressure of water is too low.
(B) Dehydrated patients may have low salt levels.
(C) Saline solutions store better.
(D) Water becomes contaminated too easily.

460. Many substances dissolve in water because they can hydrogen bond to water. Which of the following *cannot* hydrogen bond to water?

(A) CH_3OH
(B) CH_3Cl
(C) $C_6H_{12}O_6$
(D) CH_3NH_2

461. Which of the following reactions will proceed at the greatest rate?

(A) $A + B \rightarrow C$ $E_a = 100$ kJ
(B) $D + E \rightarrow F + G$ $E_a = 150$ kJ
(C) $H \rightarrow I + J$ $E_a = 250$ kJ
(D) It is impossible to determine.

462. A kinetic study of the following reaction led to a proposed mechanism:

$$Cl_2(g) + H_2O(g) \leftrightarrows HCl(g) + HClO(g)$$

The proposed mechanism was:

$Cl_2 \leftrightarrows 2\ Cl\bullet$	fast
$Cl\bullet + H_2O \rightarrow HCl + \bullet OH$	slow
$Cl\bullet + \bullet OH \rightarrow HClO$	fast

The first step is promoted by light. Which of the following are intermediates?

(A) $Cl\bullet$
(B) $\bullet OH$
(C) $Cl\bullet$ and $\bullet OH$
(D) There is no intermediate.

463. One of the important antitumor agents is cisplatin. The formula for the compound is $Pt(NH_3)_2Cl_2$. The formation of this drug is an equilibrium process. Which of the following would increase the yield of this drug?

(A) Adding $CHCl_3$ to increase the equilibrium chloride concentration
(B) Adding hydroxide ion to reduce the equilibrium platinum concentration
(C) Adding acid to react with the ammonia (a base)
(D) Increasing the ammonia concentration of the equilibrium mixture

464. Ammonium perchlorate, NH_4ClO_4, is a sparingly soluble compound. Which of the following would increase the solubility of this compound?

(A) Lowering the pH
(B) Increasing the pH
(C) Lowering the temperature
(D) Increasing the pressure

465. If the reaction quotient exceeds the K_{sp}:

(A) A precipitate will form
(B) A precipitate will dissolve
(C) The system will be in equilibrium
(D) The results depend upon the temperature

466. The addition of a catalyst to a reaction:

(A) Decreases the equilibrium constant
(B) Increases the equilibrium constant
(C) Does not change the equilibrium constant
(D) It is impossible to determine the effect on the equilibrium constant.

467. Which of the following reactions is exothermic?

(A) $A + B \rightarrow C$ $E_a = 100$ kJ
(B) $D + E \rightarrow F + G$ $E_a = -150$ kJ
(C) $H \rightarrow I + J$ $E_a = 250$ kJ
(D) It is impossible to determine.

468. The addition of a catalyst to a reaction:

(A) Does not change the heat of reaction
(B) Increases the heat of reaction
(C) Decreases the heat of reaction
(D) It is impossible to determine the effect on the heat of reaction.

469. The heat of reaction depends upon:

(A) The reactants and the transition state
(B) The reactants, the transition state, and the products
(C) The transition state and the products
(D) The reactants and the products

470. The basic reaction of respiration is:

$$C_6H_{12}O_6(aq) + 6\ O_2(g) \rightarrow 6\ CO_2(g) + 6\ H_2O(l)$$

This is a:
(A) Redox process
(B) Reduction process
(C) Oxidation process
(D) Combination process

471. Normal chemical reactions continue until the limiting reagent completely reacts. However, batteries stop producing electricity while there is still limiting reagent remaining. Why is this true?

(A) As the concentrations decrease, the reaction rate decreases.
(B) The system has reached equilibrium.
(C) The reaction will not go to completion at room temperature.
(D) The presence of the products short-circuits the battery.

472. Cytochromes are a biologically important group of iron-containing molecules. They are important in redox processes. Part of the electron transfer process in the human body involves a transfer of an electron from cytochrome to cytochrome to cytochrome. The following are the standard reduction potentials for several cytochromes:

	E°
Cytochrome b-Fe^{3+} + e^- ⇆ Cytochrome b-Fe^{2+}	+0.08 V
Cytochrome c-Fe^{3+} + e^- ⇆ Cytochrome c-Fe^{2+}	+0.22 V
Cytochrome a-Fe^{3+} + e^- ⇆ Cytochrome a-Fe^{2+}	+0.29 V
Cytochrome a_3-Fe^{3+} + e^- ⇆ Cytochrome a_3-Fe^{2+}	+0.38 V

If each electron transfer is spontaneous, what is the sequence of electron transfer for the cytochromes in the table?

(A) Cytochrome a_3 to cytochrome a to cytochrome c to cytochrome b
(B) Cytochrome b to cytochrome c to cytochrome a to cytochrome a_3
(C) Cytochrome a_3 to cytochrome c to cytochrome b to cytochrome a
(D) Cytochrome a_3 to cytochrome a to cytochrome b to cytochrome c

473. The theoretical yield of NaCl for the following reaction depends upon which of the following?

$$2\ HCl(aq) + Na_2CO_3(aq) \rightarrow 2\ NaCl(aq) + CO_2(g) + H_2O(l)$$

(A) The amount of the limiting reagent
(B) The concentration of the excess reagent
(C) The amount of NaCl recovered
(D) The temperature

474. During nuclear fission, an atomic nucleus splits into two or more smaller nuclei and possibly other small particles. This process releases large amounts of energy. The fission is random, with not all nuclei splitting in the same way. It is possible to induce the fission of a uranium-235 atom with the absorption of a high-energy neutron. If one uranium-235 splits into a cesium-140 nucleus and releases three neutrons, what other nucleus will form?

(A) Rubidium-93
(B) Rubidium-92
(C) Strontium-90
(D) Strontium-92

475. Gallium is important in the semiconductor industry as a dopant in germanium. The ground-state electron configuration of a gallium atom is:

(A) $[Ar]4s^23d^{10}4p^1$
(B) $[Ar]4s^23d^94p^2$
(C) $[Ar]4s^23d^64p^5$
(D) $[Ar]3d^{10}4p^3$

476. Mirror nuclei are nuclei in which one has the same number of neutrons as the other has protons and the same number of protons as the other has neutrons. Which of the following pairs are mirror nuclei?

(A) ^{27}Al and ^{27}Si
(B) ^{23}Na and ^{24}Mg
(C) ^{48}Ti and ^{48}Ca
(D) ^{32}P and ^{36}Cl

477. Which of the following is an excited-state electron configuration of sodium?

(A) $1s^22s^22p^63p^1$
(B) $1s^22s^22p^63s^1$
(C) $1s^22s^22p^6$
(D) $1s^22s^22p^53s^1$

478. Ionization energy increases from left to right on the periodic table. Which other trends also occur in this direction?

(A) Increasing electronegativity and decreasing atomic radii
(B) Decreasing electronegativity and decreasing atomic radii
(C) Increasing electronegativity and increasing atomic radii
(D) Decreasing electronegativity and increasing atomic radii

479. Which of the following is *not* a polar molecule?

(A) HCl
(B) CH_3Cl
(C) H_2S
(D) CF_4

480. The *best* description of the orbital geometry of an sp hybridized carbon atom is:

(A) Tetrahedral
(B) Linear
(C) Trigonal planar
(D) Trigonal pyramid

481. Which of the following elements is the *least* reactive?

(A) Ca
(B) Be
(C) Sr
(D) Ra

482. Which of the following ionic compounds will have the lowest melting point?

(A) RbI
(B) AlN
(C) CaO
(D) LiF

483. A gas mixture with a total pressure of 1,000 torr at 25°C contains 3 moles of O_2, 1 mole of H_2, and 4 moles of N_2. What is the partial pressure of H_2?

(A) 500 torr
(B) 1,000 torr
(C) 125 torr
(D) It is impossible to determine without knowing the volume of the system.

484. The ratio of the rate of effusion of methane gas is what multiple of the rate of effusion of oxygen gas?

(A) 1.4
(B) 0.25
(C) 1
(D) 32/16

485. A sample of helium gas has a volume of 500 cm³ at 227°C. What is the volume at 727°C? Assume that the pressure remains constant and that helium is nearly ideal under these conditions.

(A) 1,000 cm³
(B) 500 cm³
(C) 250 cm³
(D) 750 cm³

486. A "perfect" gas is an ideal gas, which is a theoretical substance; however, many substances approach ideal behavior under certain circumstances. Water vapor approaches ideal behavior when:

(A) The pressure is low and the temperature is high
(B) The pressure is high and the temperature is high
(C) The pressure is low and the temperature is low
(D) The pressure is high and the temperature is low

487. $Ce(NO_3)_3$ is soluble in water. Approximately how many moles of $Ce(NO_3)_3$ are necessary to produce 1.0 mole of ions in solution?

(A) 0.08
(B) 1.0
(C) 0.25
(D) 4.0

488. The following reaction obeys the rate law Rate $= k [Br_2][NO]^2$:

$$Br_2(g) + 2 NO(g) \rightarrow 2 NOBr(g)$$

Which of the following might be the mechanism for this reaction?

(A) $Br_2(g) + NO(g) \rightarrow NOBr_2(g)$ (fast)
    $NOBr_2(g) + NO(g) \rightarrow 2 NOBr(g)$ (slow)
(B) $Br_2(g) + 2 NO(g) \rightarrow 2 NOBr(g)$
(C) $Br_2(g) + NO(g) \rightarrow NOBr_2(g)$ (slow)
    $NOBr_2(g) + NO(g) \rightarrow 2 NOBr(g)$ (fast)
(D) $Br_2(g) \rightarrow 2 Br(g)$ (slow)
    $2 [Br(g) + NO(g) \rightarrow NOBr(g)]$ (fast)

489. The following equilibrium occurs at a certain temperature:

$$2 NO_2(g) \leftrightarrows 2 NO(g) + O_2(g)$$

After equilibrium is established, there is a decrease in the volume of the system, causing the:

(A) Reaction to shift to the left
(B) Reaction to shift to the right
(C) Equilibrium constant to decrease
(D) Equilibrium constant to increase

490. Which of the following acids would be the *best* choice for preparing a buffer solution of 8.6 at 25°C?

(A) $HBrO(aq)$ ($pK_a = 8.60$)
(B) $HClO(aq)$ ($pK_a = 7.30$)
(C) $HF(aq)$ ($pK_a = 3.17$)
(D) $HIO(aq)$ ($pK_a = 10.64$)

491. The conjugate bases of $H_2PO_4^-$, CH_3OH, and NH_3, respectively, are:

(A) PO_4^{3-}, CH_3O^-, and NH_2^-
(B) HPO_4^{2-}, CH_3O^-, and NH_2^{2-}
(C) HPO_4^{2-}, CH_3O^-, and NH_2^-
(D) H_3PO_4, $CH_3OH_2^+$, and NH_4^+

492. The K_a of phenol is 1×10^{-10}. Which of the following calculations will give the pK_b of the phenolate ion?

(A) $-14 + \log (1 \times 10^{-10})$
(B) $14 - \log (1 \times 10^{-10})$
(C) $14 + \log (1 \times 10^{-10})$
(D) $-14 - \log (1 \times 10^{-10})$

493. The K_a for HNO_2 is 5.1×10^{-4}. What is the approximate pK_a?

(A) 3.3
(B) 4.1
(C) 5.0
(D) 1.0

494. The following reaction will form phosgene from carbon monoxide and chlorine:

$$CO(g) + Cl_2(g) \rightarrow COCl_2(g)$$

Given the following thermodynamic data:

Compound	$\Delta H°_f$ (kJ/mol)	$\Delta G°_f$ (kJ/mol)	S° (J/mol·K)
CO(g)	−110	−137	198
$Cl_2(g)$	0	0	223
$COCl_2(g)$	−223	−210	290

The entropy change for the formation of 1 mole of phosgene is:

(A) −131 J/K
(B) −290 J/K
(C) +290 J/K
(D) −103 J/K

495. Given the following reaction:

$$2\ Al(s) + 6\ H^+(aq) \rightarrow 2\ Al^{3+}(aq) + 3\ H_2(g)$$

What will happen as the reaction proceeds?

(A) The pH will increase.
(B) The pH will decrease.
(C) The [Al] will increase.
(D) The pH will remain constant.

496. A container half-filled with ethanol, C_2H_5OH, is connected to a vacuum pump, and the air is removed from the container. How does this affect the ethanol?

(A) More ethanol molecules dissociate.
(B) The vapor pressure of ethanol increases.
(C) The boiling point of ethanol decreases.
(D) The ethanol decomposes.

497. Chlorine dioxide is an industrial bleach and a free radical. The Lewis structure for chlorine dioxide is:

(A)

(B)

(C)

(D)

498. In the presence of a catalyst, the following important industrial reaction occurs:

$$CO_2(g) + 2\ H_2(g) \rightarrow CH_2O(g) + H_2O(g)$$

What are the geometries on the carbon compounds?

(A) CO_2 is bent, and CH_2O is trigonal pyramidal.
(B) CO_2 is linear, and CH_2O is trigonal pyramidal.
(C) CO_2 is bent, and CH_2O is trigonal planar.
(D) CO_2 is linear, and CH_2O is trigonal planar.

499. How many half-lives have passed if only 6.25% of a certain radioactive isotope remains?

(A) 6
(B) 2
(C) 3
(D) 4

500. Phosphoric acid (H_3PO_4) is a triprotic weak acid, which leads to its having three K_a values. The approximate K_a values for phosphoric acid are $K_{a1} = 8 \times 10^{-3}$, $K_{a2} = 6 \times 10^{-8}$, and $K_{a3} = 4 \times 10^{-13}$. The approximate value of pK_{a1} is:

(A) 7.0
(B) 7.2
(C) 12.5
(D) 2.1

ANSWERS

Chapter 1: The Basics

1. (C) From the balanced equation, you can see that decomposition of 2 moles of copper(II) nitrate will yield 4 moles of NO_2, so that 4 mol × 46 g/mol = 184 g, about 180 g, will be formed.

2. (A) The reaction is not a neutralization (no traditional acid or base is involved with the pH approaching 7), nor is it a single replacement, since that would involve solid iron replacing the hydrogen. The oxidation states of the elements involved in the reaction have not changed, so it is not an oxidation-reduction; therefore, it is a Lewis acid-base reaction.

3. (C) The reaction releases hydrogen ions; therefore, the pH should be lower than before the reaction took place. The pH of the final solution was 5, so the pH was initially above 5.

4. (A) At high temperatures, both calcium carbonate and sodium carbonate decompose, leaving the respective oxides and releasing carbon dioxide. Na_2O is not a gas, and both oxygen and ozone would be very reactive at these temperatures, eliminating those choices.

5. (A) According to the reaction stoichiometry, 3 moles of H_2SO_4, 2 moles of $KMnO_4$, and 5 moles of $H_2C_2O_4$ will react, leaving 1 mole of sulfuric acid. The molar mass of sulfuric acid is 98 g/mol, so 1 mole will be about 100 g.

6. (D) For every 2 moles of Ag_2O that decomposes, you get 1 mole of oxygen gas; 3.0 moles will yield 1.5 moles of oxygen gas.

7. (D) AgI is an ionic compound, composed of Ag^+ and I^-. There has been no change in the iodide ion, but silver has gone from Ag^+ to solid silver by gaining an electron. The gain of electron(s) is reduction. You have no electron pair donors or receivers, so that eliminates the Lewis acid-base reaction, and nothing has become an ion that wasn't already an ion.

8. (D) A mole of H_2SO_4 would contain 2 moles of hydrogen; 5 moles of sulfuric acid would contain 10 moles of hydrogen. Hydrogen has a molar mass of about 1 g/mol; 10 moles would have a mass of 10 g. It is important to know the formulas of common chemicals.

9. (B) One mole of phosphoric acid, H_3PO_4, has an approximate mass of 98 g/mol. In 1 mole of phosphoric acid (approximately 100 g), you will have 3 moles of H (3 g or ≈ 3%), 1 mole of S (31 g or ≈ 31%), and 4 moles of O (4 moles × 16 g/mol = 64 g or ≈ 64%). Answer B is the closest.

10. (A) The sum of all oxidation numbers in the compound must equal zero, with common oxidation states for cesium being +1 and for gold being +1 and +3. However, since gold is a transition metal, other choices are possible. Therefore, (+1) + Au = 0; the oxidation state of gold is −1.

11. (A) Sulfur has an oxidation state of −2, so since there is one copper atom in answer A, it must be in the +2 oxidation state. Good test-taking strategies would indicate that since there is no answer "A and B," etc., answer A would be the only answer having no +1 species, and there is no reason to calculate the others. However, B contains Cu^{+1}, C contains Cl^{+1}, and D contains Hg^{+1}.

12. (B) The formula mass of pyruvic acid is 88.1 g/mol. In 1 mole of pyruvic acid there are 3 moles of oxygen, or 48 g of oxygen. The percentage oxygen would be (48 g/88.1 g) × 100% = 55%.

13. (C) The formula of ammonium sulfate is $(NH_4)_2SO_4$. For every mole of ammonium sulfate, there are 2 moles of ammonium ions. The solution is 0.5 M in ammonium sulfate. The concentration of ammonium ions would be two times that concentration, or 1.0 M. The ammonia will make an insignificant change in this value.

14. (D) The net ionic reaction is $OH^- + H^+ \rightarrow H_2O$. The concentration of the hydroxide ion in the sodium hydroxide solution is 0.40 M, while the concentration of the hydronium ion in the sulfuric acid solution is 2 × 0.20 M = 0.40 M. The OH^- and the H^+ are reacting in a 1:1 ratio, and since the OH^- and H^+ concentrations are the same, equal volumes are necessary.

15. (A) This is a good problem for using estimations. Eleven moles of oxygen would have a mass of 176 g [(10 × 16) + 16], while the 12 moles of carbon and 22 moles of hydrogen would have a mass of 166 g [(12 × 12) + 22]. Those are close enough to being equal to determine that the percent oxygen is slightly above 50%. The actual percentage is 176 g/342.3 g = 51.4% oxygen.

16. (A) Molarity is moles of solute per liter of solution. A density of 1 g/mL would also be 1,000 g/L. The molar mass of water is 18 g/mol. The number of moles of water in 1,000 g of water is (1,000 g)/(18 g/mol) = 55 mol. 55 mol/1 L = 55 M.

17. (A) Since this is a neutral compound, the sum of the oxidation states must be 0. Hydrogen is +1 and oxygen is −2, so for HCHO it will be (+1) + C + (+1) + (−2) = 0. Therefore, carbon is 0.

18. (D) In the perxenate ion, the sum of all oxidation states must be −4 (the charge on the polyatomic ion). Oxygen is −2, so for XeO_6^{4-}, it will be (Xe) + (6 × −2) = −4; Xe is +8.

19. (A) Reason through using approximations without working through all of the math. The molar masses of Re_2O_7 and H_2 are 484 g/mol and 2 g/mol, respectively. Therefore, 1.0 kg for each amounts to a little over 2 moles of Re_2O_7 and 500 moles of H_2. After comparing the amounts given, it is obvious that $Re_2O_7(s)$ is the limiting agent. Slightly over 2 moles of Re_2O_7 should produce slightly over 4 moles of Re. The amount of Re formed should be a little over 4 × 186 g = 744 g (= 0.744 kg).

20. (C) The sulfuric acid solution, since it is diprotic, can provide (0.100 L × 6 mol/L H$^+$) 0.6 mol H$^+$, and the potassium hydroxide solution can provide (0.200 L × 3 mol/L OH$^-$) 0.6 mol OH$^-$. There is complete neutralization, so that remaining in solution is 0.3 mol SO$_4^{2-}$ + 0.6 mol K$^+$ = 0.3 mol K$_2$SO$_4$ contained in 300 mL. 0.3 mol K$_2$SO$_4$/0.300 L = 1 M.

21. (B) Chlorine is in the +7 oxidation state in Cl$_2$O$_7$ [(2 × Cl) + (7 × −2) = 0]. In HCl, Cl is in the −1; in HClO$_3$, Cl is in the +5; and in Cl$_2$, Cl is in the 0 state. In HClO$_4$, Cl is in the [+1 + Cl + (4 × −2) = 0] +7 oxidation state.

22. (A) Oxygen is in the −2 oxidation state, and there are 4 of them, giving −8. The three irons must share a total oxidation state of +8, making each iron a +8/3.

23. (C) The empirical formula has the elements in the compound in the lowest whole number ratio. With a molecular formula of C$_6$H$_{12}$O$_6$, dividing each subscript by 6 leaves CH$_2$O. The reaction for the formation of maltose from two glucose molecules via a condensation reaction is:

$$2\ C_6H_{12}O_6 \rightarrow C_{12}H_{22}O_{11} + H_2O$$

It is not possible to simplify the formula for maltose; therefore, the molecular and empirical formulas are the same.

24. (D) $\dfrac{0.20\ \text{mol HCl}}{L} \times \dfrac{0.05000\ L}{1} \times \dfrac{1\ \text{mole Ba(OH)}_2}{2\ \text{mol HCl}} \times \dfrac{1\ L}{0.20\ \text{mol Ba(OH)}_2}$

$$= 0.025\ L\ (25\ \text{mL})$$

25. (A) There are (0.250 L × 0.10 mol/L) 0.0250 moles of OH$^-$ added to (0.250 L × 0.20 mol/L) 0.050 moles of H$^+$. The acid-base reaction occurs in a 1:1 ratio, leaving 0.025 moles H$^+$ contained in 500.0 mL. The concentration of H$^+$ would be 0.025 mol/0.500L = 0.050 M.

26. (B) The molar mass of C$_2$H$_5$OH is 46 g/mol. There are 6.02 × 10^{23} molecules per mole. 3.01 × 10^{23} molecules, therefore, would be 0.5 moles. 46 g/mol × 0.5 moles = 23 g.

27. (C) Molality is moles solute per kg of solvent. The moles of the solute would be 16 g / 32 g/mol = 0.5 moles. Molality would be 0.5 mol / 0.500 kg = 1.0 m.

28. (B) The sum of the oxidation states of all the elements in a compound will be 0. Hydrogen is +1 and oxygen is −2, so carbon must be 0.

29. (C) A mole of P$_4$ would have approximately a mass of 120 g (4 × 30 g/mol). Therefore, 1.5 moles would have a mass of 180 g.

30. (A) Assume 100 g of the compound, and convert to approximate number of moles using rounded-off masses and molar masses: Ag: 80 g / 100 g/mol = 0.8 moles Ag; P: 8 g/ 32 g/mol = 0.25 mole P; 16 g O / 16 g/mol = 1 mole O. Divide each number of moles by the smallest (0.25), giving a Ag/P/O ratio of 3.2/1/4, which is closest to answer A.

31. (A) In all compounds, the sum of the oxidation numbers must equal zero; the common oxidation numbers of Ca, H, and O are +2, +1, and −2, respectively. In $NaBrO_3$ you would have $(+1) + Br + (3 \times -2) = 0$; the oxidation number of Br would be +5. In $HBrO_2$ you would have $(+1) + Br + (2 \times -2) = 0$; the oxidation number of Br would be +3. In $Ca(BrO)_2$ you would have $(+2) + 2\ Br + (2 \times -2) = 0$; the oxidation number of Br would be +1. In $HBrO_4$ you would have $(+1) + Cl + (4 \times -2) = 0$; the oxidation number of Br would be +7. Therefore, the order is: $Ca(BrO)_2$, $HBrO_2$, $NaBrO_3$, and $HBrO_4$.

32. (B) The formula mass of ethyl alcohol is 46.1 g/mol. In 1 mole of ethyl alcohol (46.1 g), there are 2 moles of carbon (24.01 g). The percentage carbon would then be (24.01 g/46.1 g) × 100% = 52% carbon. This eliminates answers C and D. The percent oxygen would be (16.0 g / 46.1 g) × 100% = 35%, eliminating answer A.

33. (A) Formal charge of an atom = # valence electrons − (# of unshared valence electrons + 1/2 # of shared valence electrons). The cyanide ion has the Lewis structure of $:C:::N:^-$. Therefore, the formal charge on carbon is $4 − (2 + 3) = 4 − 5 = −1$; the formal charge on nitrogen $= 5 − (2 + 3) = 5 − 5 = 0$.

34. (C) Molarity is moles of solute per liter of solution. $100.0\ cm^3$ is 100.0 mL = 0.1000 L. Therefore, the molarity would be 0.2 moles solute / 0.1000 L = 2.0 M.

35. (D) Molality is moles of solute per kilogram of solvent. There is 1,000 g of the water solvent; that is 1 kg solvent. 5 moles ethanol / 1 kg solvent is a 5 m solution.

36. (A) Since there are 2 moles of sodium per mole of sodium carbonate, the original sodium ion concentration in the first solution would be 0.20 M. The actual number of moles of sodium would be 0.0500 L × 0.20 mol/L = 0.01 moles. The moles of sodium ion in the sodium chloride solution would be 0.0500 L × 0.10 mol/L = 0.005 moles. The two solutions are mixed. There are now 0.015 moles of Na^+ contained in 100 mL of solution. The molar concentration would therefore be 0.015 mol / 0.100 L = 0.15 M.

37. (B) The formula of water is H_2O with a molar mass of 18 g/mol. An Avogadro's number of particles is 1 mole of particles, and in the case of water, it would have a mass of 18 g.

38. (C) Since amounts of both reactants were given; you must first determine which is the limiting reactant. This can be accomplished by dividing the moles of reactants by their respective coefficients in the balanced chemical equation. For the HCl it would be 2.0 mol/2 = 1, and for the FeS it would be 0.5 mol/1 = 0.5. Therefore, the FeS is the limiting reactant, since it has the smaller mole-to-coefficient ratio. Now, basing the stoichiometry on the FeS, you see that for every mole of FeS that reacts, 1 mole of H_2S is formed. 0.5 moles of FeS react, forming 0.5 moles of H_2S.

Chapter 2: Atomic Structure

39. (C) 40 days = 5 half-lives (40 days/8 days per half-life). The half-life decay sequence goes 50%, 25%, 12.5%, 6.25%, and 3.13% at 5 half-lives.

40. (B) The sample would decay like this: 50% → 25% → 12.5% → 6.25% → 3.125%, and so on. The closest to 5% is the 6.25%, 4 half-lives. It would actually take a little longer than 4 half-lives to reach the 5%. Four half-lives are 32 minutes (4 × 8 min.), so the 35 minutes is the closest.

41. (B) Starting with 1 g, the decay sequence would be 1 g → 0.5 g → 0.25 g → and so on. The 0.23 g is closest to two half-lives or 2 × 5,700 years = 11,400 yrs.

42. (D) Beta particles are very penetrating and can do significant damage. Electron capture gives off x-rays, which can be very damaging. Alpha particles are not very penetrating, so answers A and D are possible. However, answer D has a very long half-life, so only a small amount of radiation is released per unit of time.

43. (C) The amount of radioactivity would reduce at the following rate: 1 $t_{1/2}$ (3.8 d) = 5×10^{-6}; 2 $t_{1/2}$ (7.6 d) = 2.5×10^{-6}; 3 $t_{1/2}$ (11.4 d) = 1.25×10^{-6}; 4 $t_{1/2}$ (15.2 d) = 6.25×10^{-7}. The radiation level will fall below 1×10^{-6} a little after 11.4 days, so 12 days is the most reasonable.

44. (A) In one half-life you would have 50% remaining, but 8 years is a little over half of a half-life, so the correct answer must be greater than 50%. This eliminates answers B and D. Radioisotopes decay as the log function, not a linear function, so you should have less than 75% remaining in 8 years. Answer A is the only one that fits this criterion.

45. (C) Thorium has 90 protons, and alpha particles have 2. The combination is 92 protons (uranium). The masses are 232 plus 4 minus the neutron loss, which comes to 235. This is ^{235}U.

46. (B) Since there is no change in the mass number, the decay mode must be one of the following: electron capture, β^- decay, β^+ decay, or gamma emission. Alpha decay reduces the mass by 4 units. Gamma emission will not change the element. Changing from xenon to iodine involves a decrease in the atomic number, which occurs in electron capture and β^+ decay. Only positron emission is listed in the choices.

47. (B) In electron capture, a proton combines with an electron to form a neutron. The atomic mass remains constant (eliminating answers C and D), and the atomic number decreases by 1. Pd is atomic number 46; therefore, the isotope with atomic number 45 is produced, Rh.

48. (D) In the alpha decay process, two protons and two neutrons are emitted. Therefore, the atomic number decreases by two. In order for the same element to be produced at the end, two protons must be created. In beta decay, a neutron is converted into a proton and an electron (beta particle). It will take two beta decays to produce the two protons needed.

49. (A) In beta emission, a neutron is converted into a proton; the atomic number increases by 1, but the mass number remains the same. In the beta decay of carbon-14, increasing the atomic number by 1 would produce N, and it should be ^{14}N because the mass number stays the same.

50. (A) Two beta decays would increase the atomic number by 2, but losing the alpha particle would decrease it by 2—there would be no net change in atomic number. However, the mass number should decrease by 4, from 230 to 226.

51. (D) Alpha particles are composed of two protons and two neutrons, a He-4 nucleus. If they gain a couple of electrons, helium gas forms.

52. (A) An alpha particle consists of 2 protons and 2 neutrons (4 units of mass). Two alpha particles would have twice the number of units of mass, or a total of 8 units of mass.

53. (D) Thorium is atomic number 90, and lead is 82. For every alpha particle that is emitted, the atomic number decreases by 2, and for every beta particle emitted, it increases by 1. The 7 alpha decays will take the atomic number from 90 to 76, and the 6 beta decays will increase it from 76 to 82.

54. (C) Nickel is $[Ar]4s^2 3d^8$ with both 4s electrons paired and eight 3d electrons distributed amongst five 3d orbitals, leaving three pairs and two unpaired. Calcium is $[Ar]4s^2$ with no unpaired electrons. Vanadium is $[Ar]4s^2 3d^3$ with three unpaired 3d electrons. Aluminum is $[Ne]3s^2 3p^1$ with one unpaired 3p electron.

55. (C) Answers A and B refer to the actual mass of an atom, while answer D refers to the mass in grams of a mole of a substance. The mass defect refers to the difference between the mass of an atom (isotope) and the sum of the masses of its protons, neutrons, and electrons.

56. (B) The electron configuration indicates that there are 12 electrons, and the problem states that it is an element, not an ion, so that there are also 12 protons. The element with an atomic number of 12 is magnesium.

57. (A) The $ns^2 np^6$ configuration would indicate the presence of p electrons, which is true of all the noble gases except He.

58. (C) The electron configuration indicates that the element has 36 (Kr kernel) + 9 = 45 electrons. The element with atomic number 45 is Rh.

59. (D) The mass number of an element is the number of protons plus the number of neutrons present in an atom. For this reason, the mass number must be a whole number. The 17 for chlorine is the atomic number, which is the same as the number of protons. There is no number present giving the number of neutrons. The number 35.453 is the mass number of chlorine. The mass number is the weighted average mass of the isotopes of an element and is rarely a whole number. However, for some man-made elements, the mass number of the most stable isotope is sometimes given.

60. (B) An electric field will affect any species that contains a charge. All of the answers contain charged species except gamma rays. The greatest deviation will be for the lightest particle, which is the β^- particle, as it is only an electron.

61. (A) In all these compounds, the sum of the oxidation numbers must equal 0, since the compounds are neutral. Hydrogen and potassium are in the +1 state and fluorine is −1. Therefore, in answer A: O + (2 × −1) = 0, so O is +2; in answer B: O + (2 × +1) = 0, so O = −2; in answer C, 2 O + (2 × +1) = 0, so O = −1; and in answer D: (+1) + 2 O = 0, so that O = −½. Answer A is the highest.

62. (C) Fe_2O_3 contains Fe^{3+}, which has 23 electrons. Cr^{2+} has 22 electrons, Co^{2+} has 25 electrons, Mn^{2+} has 23 electrons, and Fe^{2+} has 24 electrons.

63. (D) In the n = 4 shell, you will have the 4s (2 electrons), the 4p (6 electrons), the 4d (10 electrons), and the 4f (14 electrons). That gives 32 electrons.

64. (C) For bohrium, $n = 7$, l values will range from 6 to 0. The m_l values will be the ones that contain a quantum number of −3. The first orbital for which this value is possible is an f orbital. The only f orbital listed is the 5f orbital.

65. (A) *Degenerate* is a term that refers to orbitals that have the same energy.

66. (A) In beta emission, a neutron is converted into a proton and a beta particle that is expelled from the nucleus. Therefore, the atomic number of the product is one unit higher than that of the particle that underwent beta emission. Iodine has an atomic number of 53. The product would have an atomic number of 54, which is Xe.

67. (A) The Bohr model worked well for atoms or ions that have a single electron. Those that have a single electron are Bohr atoms. He^+ has lost one of its two electrons, leaving one. H^- has two electrons, and H^+ has none. Li^+ forms when lithium loses its valence electrons, leaving two.

68. (D) ^{15}N has a magic number of neutrons (8). None of the others have a magic number of either protons or neutrons.

69. (C) Fe (Z = 26) does not have a magic number of protons and is therefore eliminated. The remaining answers have magic numbers of protons. Sn-118 would have (118 − 50) 68 neutrons and is eliminated. O-17 has (17 − 8) 9 neutrons and is eliminated. Pb-208 would have (208 − 82) 126 neutrons, which is a magic number, and so answer C is correct.

70. (C) Xe-128 has (128 − 54) 74 neutrons, while Ba-130 has (130 − 56) 74 neutrons; they are isodiaphers, while none of the other pairs have the same number of neutrons.

71. (D) In beta emission, a neutron is converted into a proton with emission of a beta particle, thus reducing the number of protons and increasing the number of protons. This decreases a high neutron/proton ratio.

72. (A) The flame is emitting brilliant crimson light. Thus, energy is emitted. As the electrons drop from a higher energy state to a lower one (for example, 4p → 2s), energy is released (this eliminates answers C and D). This drop results in the emission of a bright band in the atomic spectrum.

73. (A) K is $[Ar]4s^1$, while all the others are [Ar].

74. (D) Cobalt atoms have 27 electrons, while Co^{3+} cations have 24 electrons. All answers use the [Ar] core (18 electrons), so the ion has an additional 6 electrons ($18 + 6 = 24$). That would be answers B and D. However, the 4s electrons always leave before the 3d electrons, so the result will have no 4s electrons.

75. (C) The attraction to a magnet is due to paramagnetism. Paramagnetism is a consequence of unpaired electrons. Answers A and D have an unpaired s-electron, while answer B has an unpaired d-electron. In the Cu^+, there are only paired electrons; hence it is diamagnetic, not paramagnetic.

76. (A) Silicon has 14 electrons. (This eliminates all the answers except answer A, because the others show 15 electrons.) The electron configuration of silicon in its ground state would be $1s^2 2s^2 2p^6 3s^2 3p^2$. It would be one of the 3p electrons that would be promoted to a higher energy state, such as the 4s.

77. (B) The amount of energy produced can be calculated by Einstein's equation, $E = mc^2$. The speed of light, c, is 3.0×10^8 m/s. The units of J = kg·m/s, so the mass must be expressed in kg, 2.0×10^{-31} kg. Therefore, $E = (2.0 \times 10^{-31})(3.0 \times 10^8)^2 = 1.8 \times 10^{-14}$ J.

78. (C) As the two nuclei fuse, energy is released. That energy comes from a small amount of mass being converted into energy. Therefore, the nucleus formed would have a mass slightly smaller than the sum of the two original nuclei.

79. (D) Sodium has a single valence electron in the 3s orbital. (This eliminates answer C, where the energy level is shown as 2.) The valence electron is in an s orbital ($l = 0$), so answer A is eliminated. The only value of m_l with $l = 0$ would be 0, eliminating answer B.

80. (D) No two electrons may have the exact same set of four quantum numbers.

81. (B) Sp^2 hybridized carbons are those involved in double bonds. In the fused ring system, there are 5 carbons that are part of a double bond.

82. (C) Transition metal compounds are often colored. In most cases, this color is due to partially filled d orbitals. Fe_2O_3 is commonly called rust and is certainly colored. It is the only element with a partially filled d subshell.

83. (D) Taking a diuretic would not decrease the iodine-131, since it tends to concentrate in the thyroid and would eliminate both isotopes. Boiling will not affect the rate of decay of any radioisotope. Adding silver nitrate would precipitate both iodine isotopes. The only choice that would work is to increase the intake of nonradioactive iodine, thus diluting the radioactive form. That would increase the possibility of nonradioactive iodine being absorbed.

Chapter 3: Bonding and the Periodic Table

84. (A) There should be 12 electrons (6 from each oxygen) in the Lewis structure of diatomic oxygen. Only structure A has 12 electrons being used, with each oxygen having a full octet. (Counting electrons and bonds is important while taking the MCAT.)

85. (C) This binding does not involve a redox reaction, eliminating answers A and D. The oxygen molecule has several pairs of valence electron that can be donated to the iron. This makes the oxygen a Lewis base, donating a pair of electrons.

86. (D) In the bonding of oxygen to the iron in hemoglobin, the oxygen acts as a Lewis base, furnishing a lone pair of electrons for a coordinate covalent bond.

87. (D) Trigonal bipyramidal has five groups around the central atom; square planar has four groups and two lone pairs; tetrahedral has four groups; and octahedral has six groups and no lone pairs.

88. (A) Residues refer to the group (side chain) attached to the amino acid backbone-amine-carbon-carboxylic acid. In answers B, C, and D, the residues are either carbon or hydrogen atoms that have no nonbonding electron pairs. However, aspartic acid has a residue that contains two oxygen atoms that have nonbonding electron pairs that could form coordinate covalent bonds with an iron ion.

89. (C) Potassium and arsenic are in the same period. The atomic radius of an atom decreases as you move from left to right in a period because the increasing effective nuclear charge pulls the valence electrons closer to the nucleus. Therefore, potassium would be larger than arsenic due to the smaller effective nuclear charge.

90. (A) Magnesium chloride is an ionic compound (metal + nonmetal), and therefore no electrons are shared—they have been transferred.

91. (A) The third ionization energy of calcium will be much higher than that of scandium since in calcium you have removed the 4s electrons and are now trying to remove a 3p electron (core electron) that is much closer to the nucleus. With scandium, the three electrons are in the same energy level and do not require as much energy.

92. (B) The emission of light is due to electron(s) dropping from a higher energy state to a lower one, since they must release energy. To reach an excited state, energy must be absorbed; that energy is released when those electrons drop back down to their ground state. Each transition would have a certain energy (quantum) and a certain wavelength associated with it.

93. (A) Helium atoms are capable of forming London dispersion forces, which are due to a temporary dipole-dipole attraction.

94. (A) All the species are isoelectronic; therefore, they all begin with the same size. However, the charge will make a difference. Positive charges decrease the radii, and negative charges increase the radii.

95. (C) Since the speed is inversely proportional to the ionic radius, the slowest (smallest) rate of diffusion would be by the ion with the largest ionic radius. This would be the barium ion, since it is in period 6.

96. (B) Metals readily form cations. Low ionization energies allow cations to form more readily. The general trend in increasing metallic character parallels the general trend in decreasing ionization energy.

97. (D) During the first electron affinity, a negative electron is being added to a neutral atom. However, during the second electron affinity, a negative electron is being added to an anion that already has a negative charge. The repulsion of the like charges must be overcome to add the second electron, and thus the second electron affinity value is much greater than the first.

98. (B) A lower ionization energy is related to the fact that the electron being lost is farther from the nucleus. Sodium and chlorine are in the same period. Because of the increasing effective nuclear charge as you go from left to right on the periodic table, the atoms get smaller and the ionization energy increases, since the valence electrons are closer to the nucleus. The chlorine atom is smaller than the sodium atom. The valence electron in sodium is farther from the nucleus than in chlorine, and thus sodium has a smaller ionization energy.

99. (B) Formal charge (FC) = # valence electrons − (# nonbonding electrons + 1/2 # shared electrons). For the hydrogen, $1 - (0 + 1) = 0$. There is no need to calculate the other atoms, since answer B is the only one that contains a zero in the first position.

100. (D) The left oxygen has two bonds and two lone pairs in a tetrahedral orientation. That is sp^3 hybridization. The right oxygen has a double bond, which is sp^2 hybridization.

101. (A) When a hydrogen ion is lost, the nitrite ion can be written in two possible resonance forms with the nitrogen to oxygen bonds switched. Therefore, in the resonance hybrid, the bonds are equal in length, with a bond order of 1.5.

102. (C) Both carbon atoms have three groups attached. (Remember that a double bond counts as a single group.) The geometry with three groups around a central atom is trigonal planar.

103. (B) The formula of the carbonate ion is CO_3^{2-}. There are three groups surrounding the carbon atom, a resonating double bond, and no lone pairs. Therefore, the molecular geometry is trigonal planar.

104. (D) This is an example of an acid-base reaction. A hydrogen ion transfers from the HCl to the ammonia, forming ammonium cations and chloride anions, which form an ionic compound. Most ionic compounds are solids at room temperature.

105. (A) Both ethyl alcohol and carbon monoxide are molecular compounds that do not ionize in water. They are not electrolytes. HBr will ionize, producing hydrogen ions and bromide ions, and is an electrolyte.

106. (D) Free radicals contain one or more unpaired electrons. Any species with an odd number of electrons is a free radical. Otherwise, it would be necessary to draw Lewis structures or MO diagrams. All the answers have an even number of electrons except NO, which has 11 total valence electrons. Therefore, NO is the free radical. If all the answers had an even number of electrons, it would be necessary to construct molecular orbital energy level diagrams. Simple Lewis structures will not work, as, for example, the Lewis structure for the oxygen molecule incorrectly shows all electrons paired instead of the two unpaired electrons correctly shown in the MO diagram.

107. (C) All the bond angles are 120°; therefore the hybridization is sp^2 with one H attached to one of the hybrid orbitals. The radical contains seven electrons. These bonds use six of the valence electrons for the bonds. The remaining p-orbital is perpendicular to the plane and holds the last electron.

108. (C) Ammonia has tetrahedral orbital geometry because there are four electron pairs around the central nitrogen atom.

109. (B) Silicon tetrafluoride would have the same geometry as carbon tetrachloride (tetrahedral). It would be a covalent compound (nonmetal + nonmetal). There are no common reactions that would cause rapid oxidation of SiF_4, resulting in an explosion. Because it is tetrahedral, it would be nonpolar.

110. (A) In nitric acid, the hydrogen is bonded to one of the oxygen atoms. In order for oxygen to have a complete octet, there must be a double bond (4 electrons) and two single bonds (2×2), for a total of 8 electrons.

111. (D) Neither calcium ions nor ethane have nonbonding electron pairs, eliminating answers A and C. Answer B is false since the presence of the electron pairs would reduce any positive charge. A Lewis base is defined as an electron pair donor, and that is what ligands do, making answer D correct.

112. (B) The bonding between the atoms in the bicarbonate ion (intramolecular force) is best described as covalent. In the case of the bicarbonate ion, ionic bonding, hydrogen bonding, and dipole-dipole forces would be intermolecular—not intramolecular.

113. (A) Hydrogen bonding in water is a relatively strong intermolecular force. It is stronger than the hydrogen bonding that occurs in ammonia, since oxygen is more electronegative than nitrogen. Water's hydrogen bonding is stronger than the dipole-dipole forces in CH_3F and much stronger than the London dispersion forces in CH_3CH_3. It is, however, weaker than the ionic bonding that is in sodium chloride.

114. (D) The O–N–O bond angle is less than predicted because the lone pair of electrons on the nitrogen spread out slightly, forcing the bond angle to be a little smaller.

115. (A) In the Lewis structure of the ion, there are no double bonds, four single bonds, and two lone pairs on the central chlorine. This combination results in a square planar geometry. The other geometries in the problem have one or no lone pairs.

116. (B) The level is always read at the bottom of the meniscus. Many times, it is difficult to tell exactly where the top of the meniscus is located.

117. (C) Both ions are linear, and all forms contain two σ-bonds and two π-bonds. The most stable resonance structures for the cyanate ion have a -1 formal charge either on the oxygen or on the nitrogen, with a zero formal charge on the other two atoms. The best resonance structure for the fulminate ion has a -1 formal charge on both the carbon and the oxygen, with a $+1$ formal charge on the nitrogen. There are also fewer good resonance structures for the fulminate ion.

118. (D) The electromagnetic spectrum in order of decreasing energy is: gamma ray $\rightarrow$ x-ray $\rightarrow$ ultraviolet $\rightarrow$ visible $\rightarrow$ infrared $\rightarrow$ microwave $\rightarrow$ radio waves. Since the vibrational transitions require less energy, they will be to the right of the ultraviolet region. The only choice to the right is infrared.

119. (A) Xe is monatomic with no vibrational spectrum, since there are no bonds. All the others are molecules with vibrational spectra due to their bonds.

120. (D) CF_4 is the only nonpolar species in the group.

121. (B) CHF_3 is the only compound listed that does not have a hydrogen atom attached to a N, O, or F. The hydrogen bonds to carbon in this case.

122. (A) The geometry of PF_5 is trigonal bipyramidal because of the five bonds to the phosphorus and the absence of any lone pairs on the phosphorus atom. A square pyramid requires five bonds and a lone pair. These facts eliminate answers C and D. Formal charge (FC) = # valence electrons $-$ (# nonbonding electrons $+ 1/2$ # shared electrons). Phosphorus has 5 valence electrons, so it would have the following formal charge: FC = $5 - [0 + 1/2(10)] = 0$.

123. (A) While Cs has the lowest first ionization energy of this group; the second ionization energy involves a disruption of the core electrons. Hydrogen does not have a second ionization energy. Ba is larger than Ca; therefore, both the first and second ionization energies of Ba are lower than those of Ca are.

124. (D) Metals tend to be malleable, ductile, and good conductors; they are not brittle. Nonmetals tend to be brittle.

125. (A) The boron has a vacant orbital that can accept an electron pair. That makes it a Lewis acid. BX_3 would have a trigonal planar geometry (only three electron pairs around the boron), but it changes to tetrahedral when the fourth electron pair is accepted.

126. (A) Within a period, electronegativities increase as you move from left to right on the periodic table. In the F–F bond, there would be a zero polarity difference, since the electronegativities are the same. This eliminates answers C and D. The greatest will be between C and F, since the difference in electronegativities is the greatest. Therefore, answer A is correct.

127. (C) Because of the properties of diamond (hardness, etc.), the bonds between the carbons must be strong. This eliminates answers A and D because these are weak forces. The bond between Zn and S is ionic (metal + nonmetal), but the bond between carbon atoms must be covalent (nonmetal + nonmetal).

128. (B) The stronger the intermolecular force, the higher the melting point is. Water's intermolecular force is hydrogen bonding; HCl has dipole-dipole attraction, while chlorine has London forces. All of these are weak relative to the ionic bonding present in NaCl. In terms of strength, ionic bonding > hydrogen bonding > dipole-dipole > London.

129. (A) The molar mass of HCl is 36.5 g/mol. Therefore, 73 g would be 2.0 moles, but since it is a strong acid, you would get 4.0 moles of particles. Since freezing point depression is a colligative property, it just depends on the number of particles present. If 2 moles of particles lowered the freezing point 3.8°C from pure water's freezing point of 0°C, then 4 moles should lower it 7.6°C. Therefore, the freezing point of the solution should be −7.6°C.

130. (C) Ionization energy, the energy needed to remove a valence electron from an atom, increases from left to right in a period and decreases from top to bottom in a group. The element with the lowest ionization energy should then be in the leftmost group in the period and the lowest on the periodic table. That would be potassium in this case.

131. (C) Both electronegativity and electron affinity are related to the increasing effective nuclear charge, which occurs with increasing atomic number within a period. In general, as the atomic number increases, the atoms become smaller and the attractive force for the valence electrons increases (increasing electronegativity). This eliminates answers A and B. The electron affinity generally increases with increasing atomic number within a period, since the increasing effective nuclear charge should attract an electron being added to the valence shell. This eliminates answer D.

132. (A) Linear, tetrahedral, and octahedral are all high symmetry geometries and thus can be nonpolar. Bent molecules are less symmetrical and cannot be nonpolar.

133. (C) Sodium, because it is the leftmost member of its period, has a low ionization potential. It is relatively easy for it to lose an electron, forming a cation and reaching noble gas configuration.

134. (B) Argon is a noble gas and therefore is very difficult to react. All of the other choices will react with potassium.

135. (B) Answer A is false because chloride ions are much larger than sodium ions. Answer C is false since calcium ions are smaller than potassium ions. Answer D is false since potassium ions would be larger than sodium ions. Answer B is correct since the sodium ion would be smaller than the potassium ion, since potassium has added a major energy level.

136. (C) Neon is a noble gas with a full octet. It would require a tremendous amount of energy to add an additional electron. All the rest are nonmetals that readily form anions.

137. (C) The sulfide ion is P^{3-}. It has gained three electrons and now has the same number of electrons as argon, Ar.

138. (A) Formal charge (FC) = # valence electrons – (# nonbonding electrons + ½ # shared electrons). Nitrogen has 5 valence electrons, so it would have the following formal charges in the answers: answer A: FC = 5 – (2 + 3) = 0; answer B: FC = 5 – (6 + 1) = –2; answer C: FC = 5 – (4 + 2) = –1; answer D: FC = 5 – (6 + 1) = –2.

139. (A) Lowering the pH will add a hydrogen ion to the pyruvate ion, making it neutral. This will decrease its solubility in water. This eliminates answer B. In the ion shown, the carbon-oxygen bonds on the left of the diagram participate in resonance, which makes the bonds of equal length and longer than the bond to the upper right oxygen atom. This eliminates answers C and D.

140. (D) A square pyramidal geometry has five atoms and one lone pair surrounding a central atom. Both antimony pentafluoride and iodine pentafluoride have five atoms around the central atom; however, only iodine pentafluoride also has a lone pair on the central atom.

141. (A) The reaction is $SO_3(g) + H_2O(l) \rightarrow H_2SO_4(aq)$. The Lewis structure of sulfur trioxide has three single bonds to the oxygen atoms and a resonating double bond to the oxygen atoms, making the molecule trigonal planar. This eliminates answers B and C. In sulfuric acid, the hydrogen atoms are attached to oxygen atoms, making two of the oxygen atoms single bonded to the sulfur and doubled bonds to the remaining two oxygen atoms. That gives four groups around the sulfur and no lone pairs on the sulfur, which describes a tetrahedral geometry.

142. (A) Cell membranes are hydrophobic, allowing the passage of nonpolar materials more easily than polar or ionic ones. Since dimethyl mercury is nonpolar, it can pass through the cell membranes more easily than the cationic mercury.

143. (C) Melting points can be correlated with molar mass and intermolecular forces. The higher the molar mass, the higher the melting point. All molecules except answer C can participate in hydrogen bonding. The lower intermolecular attraction in answer C leads to a lower melting point. Answer C also has the lowest molar mass.

Chapter 4: Phases

144. (A) The heat of vaporization is related to the strength of the intermolecular forces in the liquid. The stronger the intermolecular forces, the higher the heat of vaporization. Water exhibits hydrogen bonding, a strong intermolecular force, between molecules. It takes a lot of energy to break these strong hydrogen bonds between the molecules, so the heat of vaporization is high.

145. (D) Fusion plus vaporization represents the process of going from a solid to a gas; this is sublimation. The heat of sublimation is the sum of these two energies. Deposition is the opposite of sublimation. The energy, with changed sign, is the same.

146. (B) 54 grams of ice are 3.0 moles of ice. To completely melt all the ice, it would take (6.0 kJ/mol × 3.0 moles) 18.0 kJ. Only 16.0 kJ are being added. Therefore, not all of the ice will melt, and the temperature will remain constant at 0°C.

147. (C) The normal boiling point is the temperature at which a substance boils under 1.00 atm pressure (P_4). Following the dotted line from P_4 shows that it intersects the liquid-gas line at T_4. Therefore, P_4 and T_4 are the pressure and temperature at the normal boiling point.

148. (C) The water is initially at T_1, and the temperature remains constant. The water vapor is compressed, which causes the pressure to increase along the vertical dotted line. It goes from a gas to a solid and then to a liquid.

149. (D) At the top of the mountain, the atmospheric pressure is lower and less heat is necessary to heat the water enough so that the water's vapor pressure is equal to the atmospheric pressure. Therefore, the water boils at a lower temperature.

150. (A) Water exhibits hydrogen bonding between the molecules. When water begins to freeze, a crystal structure is formed using those hydrogen bonds. In this crystal structure, there are large empty spaces that make the density of ice less than the density of water.

151. (A) To lower the freezing point of the water and prevent the sidewalk from cracking, a substance is added to make a solution that has a lower freezing point. This freezing point depression is a colligative property, so it really depends on the number of solute particles present. Answers B, C, and D involve salts, so for every mole of solute added, you get 2 to 4 moles of particles. Answer A, however, is a molecular substance, so for every mole added, you only get 1 mole of particles; it will have the least effect.

152. (B) At the pressure given and −10°C, the substance would be in the solid phase. Increasing the temperature at that pressure would cause the substance to move from the solid to the gaseous phase. This process is sublimation.

153. (B) The heat capacity is the specific heat of the substance times its mass. It has units of J/°C. The units eliminate answers A and C. 4.2 J/g°C × 200 = heat capacity = 840 J/°C.

154. (B) Deposition is the phase change from a gas to a solid. The transition from a gas to a solid releases energy and the entropy decreases.

155. (B) The heat of sublimation is equal to the sum of the heat of fusion plus the heat of vaporization. This fact eliminates the other choices.

156. (C) Sublimation is a process that starts with a solid and goes directly to a gas. Solidification ends with a solid. Deposition and condensation start with a gas, not a solid.

157. (D) Normal boiling point refers to the temperature at which a liquid will boil at 1 atm pressure. If the pressure is increased, the liquid will boil at a temperature greater than its normal boiling point.

158. (A) Normal boiling point refers to the temperature at which a liquid will boil at 1 atm pressure. Answer D, STP, assumes a temperature of 0°C. Answer B is wrong because changing the pressure affects the boiling point, and answer C has nothing to do with boiling point.

159. (C) Graphite is softer than diamond because there are relatively weak London dispersion forces between the sheets of graphite, while in a diamond there are strong covalent bonds throughout the structure. The stronger the bonds holding the entire structure together, the harder the material.

160. (A) The enthalpy of sublimation is related to the strength of the intermolecular forces. Hydrogen chloride and formaldehyde have dipole-dipole attraction, ethane would have London dispersion forces, and methanol would have hydrogen bonding. Hydrogen bonding is the strongest of these; therefore, ethanol would have the highest heat of vaporization.

161. (A) Answer B is incorrect because the pressure is due to a variety of gases, not just ozone. Answer C cannot be correct because the partial pressure of ozone cannot be greater than the total pressure. Answer D is incorrect because the energy of collisions is not affected by pressure, just temperature. Answer A is correct because the partial pressure of one component must be less than the total pressure (8 torr).

162. (D) In order for a liquid to freeze, energy must be lost to the surroundings. In deposition, going from a gas to a solid, energy must be lost to the surroundings. The same is true for condensation, going from a gas to a liquid. However, in sublimation, in order for a solid to go into the gaseous phase, energy must be absorbed from the surroundings.

163. (C) Heat capacity has units of J/K, while the specific heat has units of J/g·K. Multiplying the specific heat by the mass (g) of the substance generates the heat capacity.

164. (C) Answers A, B, and D represent phase transitions that would be represented as horizontal portions on the graph. The specific heats are necessary to calculate the heat needed to go from one temperature to another.

165. (A) The melting point occurs at T_2, and boiling occurs at T_3. These temperatures must be known in order to calculate the temperature change from the liquid at the melting point to the liquid at the boiling point.

166. (A) The heats of fusion of ionic compounds are normally much higher than those of covalent substances, even polar ones. The heat of fusion of aluminum oxide will be higher than that of NaCl because of the increased attraction of the 3+ cations and 2- anions versus the 1+ to 1- attraction in sodium chloride. These facts eliminate answers B, C, and D. (The heat of fusion of a polar covalent substance [H_2O] should be greater than that of a nonpolar one [CCl_4].)

167. (C) NaCl is a nonvolatile solute, so it would not appear in the gas phase. Water does not decompose upon boiling, so only water vapor is in the gas phase.

168. (A) The heat of sublimation is the heat necessary to both melt and vaporize a substance. With both graphite and diamond, extremely strong covalent bonds must be broken, requiring a large amount of energy. Answers B, C, and D are incorrect because they do not mention covalent bonds.

169. (C) This is a Hess's law problem. Looking at the desired reaction, you can see that you need 2 N_2 on the left, so you can multiply the third equation by 2 along with its ΔH (giving -700 kJ). You need 2 N_2O_5 on the right, so reverse equation two and multiply by 2 (giving +160 kJ). Finally, since you will need 2 H_2O on the left to cancel the 2 waters that are now on the right, it is necessary to reverse equation one (giving + 570 kJ). Summing up the three ΔH values, you get 30 kJ.

170. (B) Specific heat has the units of J/g•K. In this case, there was a 60.0°C change in temperature, which is also a 60.0 K change in temperature (no need to convert from °C to K). Therefore, the specific heat would be 300.0 J / (50.0 g × 60.0 K) = 0.100 J/g•K.

171. (A) The molar mass of ethanol is 46 g/mol, so that 0.46 g of ethanol is 0.01 moles. The heat absorbed by the calorimeter would be 1.2 kJ/°C × 10°C = 12 kJ, and it would be negative, since the reaction was exothermic. The heat of reaction would be −12 kJ / 0.01 moles = −1.2 × 10^3 kJ.

172. (B) The heat change would be the specific heat × mass of solution × change in temperature = 4.2 J/g•K × 158 g × 3.0 K = 1.99 × 10^3 J, or about 2 kJ. This enthalpy change was for 0.10 moles of sodium nitrate [(8.6 g) / (86 g/mol)], so the enthalpy change would then be 2 kJ / 0.10 mol = 20 kJ/mol, and the value would be positive since the reaction was endothermic.

173. (A) 260 g of KI is about 2 moles. The total energy change will be KI(s) → KI(l) + heating KI(l) from −50°C to −35°C + KI(l) → KI(g) + heating −35°C to 0°C: [2 mol × 2.9 kJ/mol] + [2 mol × 44 J/mol·K × 15 K] + [2 mol × 44 kJ/mol] + [2 mol × 25 J/mol·K × 35 K] = 97 kJ. Appropriate approximations should be applied to avoid arithmetic.

174. (C) Remember that in thermochemistry problems, it is products minus reactants. The heats of formation of the elements, potassium and hydrogen, are zero. Therefore, 2 (ΔH_f° of KOH) − 2 (ΔH_f° of H_2O) = −130 kJ. Then substitute the standard heat of formation of water and solve for the standard heat of formation of KOH, which equals −350 kJ/mol.

175. (C) The heat of vaporization is related to the strength of the intermolecular forces in the liquid. The stronger the intermolecular forces, the higher the heat of vaporization. Methanol exhibits hydrogen bonding between molecules, while methyl chloride has dipole-dipole attraction as its predominant intermolecular force. Hydrogen bonding is a stronger force than dipole-dipole attraction; so methanol has a higher heat of vaporization.

176. (B) At the boiling point, the energy needed to vaporize the liquid is the heat of vaporization times the mass of the substance. Therefore, $0.500 \text{ kJ/g} \times 100.0 \text{ g} = 50.0 \text{ kJ}$.

177. (B) The only time you can add energy to a sample and have the temperature remain constant is at a phase transition. The problem states that the ethanol is at a temperature below its boiling point. The only other phase transition would be melting, so the sample is at its melting point.

178. (B) Since energy must be absorbed to boil a liquid, the enthalpy change is positive. This eliminates answers C and D. This energy is used to break the forces between the particles, the intermolecular (not intramolecular) forces.

179. (B) Deposition is going from the gaseous phase to the solid phase: $SO_3(g) \rightarrow SO_3(s)$. It would be the difference in the enthalpies of formation of products minus reactants or $-454 \text{ kJ/mol} - (-396 \text{ kJ/mol}) = -58 \text{ kJ/mol}$.

180. (A) The first horizontal region represents melting. The second horizontal region represents vaporization: going from a liquid to a gas. This eliminates answers D and C. Normally, it takes more energy to convert a substance into a gas than to melt a solid, so that the heat of vaporization is greater than the heat of fusion.

181. (A) The enthalpy change would be the specific heat of solution times the mass (solute + water) of the solution times the change in temperature. It would be a negative value, since the process was exothermic (that eliminates answers B and D). Enthalpy change = 4.2 $\text{J/g}°\text{C} \times 200 \text{ g} \times 10.0°\text{C} = -8.4 \text{ kJ}$.

182. (D) The critical point is the point at which the gas and liquid phases are indistinguishable from each other.

Chapter 5: Gases

183. (C) Since the pressure was constant, the only effect will be the change in temperature from 273 K to 546 K ($\times 2$). The initial volume was 11.2 L, and the volume should increase with an increase in temperature. The new volume will be $11.2 \text{ L} \times 2 = 22.4 \text{ L}$.

184. (C) The piston was locked in place, so the initial and final volumes are the same, eliminating answers A and D. The temperature was doubled (273 K to 546 K), so at constant volume the pressure should double, eliminating answer B. Answer C is correct with $P_f = 2 P_i$ and the two volumes equal.

185. (A) The gas is heated to 546 K, and then the piston is unlocked. The gas will expand, $V_f > V_i$, and as it expands, the pressure will decrease, since the volume is greater, $P_f < P_i$.

186. (A) Halving the diameter of the piston will cause the volume to decrease to one-half its former volume; doubling the pressure will cause the piston to move inward, doubling its initial volume. The result of the two processes is no net change in volume.

187. (A) The system initially is at STP, so the 0.5 moles of gas would occupy 11.2 L. (1 mole of any gas at STP occupies 22.4 L.) Now you can convert the volume to cubic meters: 11.2 L × (1,000 mL/1 L) × (1 cm³/1 mL) × (1 m)³/(100 cm)³ = 1.1 × 10⁻² m³.

188. (D) Dalton's law states that the sum of all partial pressures of gases in a mixture equals the total pressure. This means that the partial pressure of a gas is the percentage (or mole fraction) of the gas times the total pressure. In this case, (35%/100%) × 1,000 torr = 350 torr.

189. (B) Using the ideal gas law, PV = nRT, where PV and R are constant. An increase in the number of moles must be compensated for by a decrease in the temperature to keep nRT constant.

190. (C) Gases that are at the same temperature have the same average kinetic energies.

191. (D) In order for a shock absorber to work, the material must be able to compress to absorb the shocks. Gases are compressible, so shock absorbers are normally filled with air or nitrogen gas. Water would not work, because it is not compressible. Answer A is incorrect since you would not be exerting enough pressure in a shock absorber; answer B would be correct, but shock absorbers must work at lower temperatures; therefore, answer D is a better reason. Answer C has nothing to do with shock absorbing ability.

192. (C) Consider the ideal gas equation, PV = nRT. Solving for P: P = nRT/V. If the pressure halves and the volume stays constant, the temperature must also halve (R is a constant, and the number of moles remains constant).

193. (D) The diffusion of gases (Graham's law) relates to the velocity (v, not V) of the gas particles. Answer A is the modification of the ideal gas law for real gases, answer B gives the pressure/volume relationship of gases, and answer C is the ideal gas equation.

194. (C) Since the cloud formed near the center, the molar mass of the unknown is close to that of dimethylamine, 45 g/mol. HI = 128 g/mol, HF = 20 g/mol, HCOOH = 46 g/mol, and HSCN = 59 g/mol. HCOOH is the closest.

195. (B) The most ideal will be the one with the weakest intermolecular forces. HF is highly polar, so it is eliminated. The remaining three have London dispersion forces. Helium is the smallest (fewest electrons); therefore, it has the weakest London dispersion forces, which leads to the minimum deviation.

196. (A) 76 torr = 0.1 atm and (10,000 Pa) / (1 × 10⁵ Pa/atm) = 0.1 atm, so the total pressure is (0.50 + 0.10 = 0.10) atm = 0.70 atm.

197. (B) Converting the masses to moles gives CH₄ = 2.0 mole; C₂H₆ = 2.0 mole; and C₄H₁₀ = 2.0 mole. The mole fraction of each, times the total pressure, will give the partial pressure of each gas. (2 mol/6 mol) × 2.40 atm = 0.80 atm for each gas.

198. (A) 1.0 mole of any gas at 1 atm and 273 K (0°C) = 22.4 L. Therefore, there would be 22.4 L × 5.8 g/L = 130 g in 1.0 mol. The molar masses are CO_2 = 44 g/mol, CH_4 = 16 g/mol, H_2 = 2.0 g/mol, and Xe = 131 g/mol. The gas must be xenon.

199. (B) The volume and Kelvin temperature of a gas are directly related at constant pressure and moles. The gas began at a temperature of 400 K. The volume was reduced by one-half; therefore, the temperature would decrease by one-half. The new temperature would be 400 K/2 = 200 K = −73°C. Answer B is the closest.

200. (C) Since the temperature is the same, so is the average kinetic energy. Since the volume, pressure, and temperature are the same, then the moles must be the same. Oxygen has a greater molar mass than hydrogen; therefore, an equal number of moles of oxygen will have a greater mass. A greater mass in the same volume will yield a greater density.

201. (A) If you have two samples of gas at the same temperature, they have the same average kinetic energies. However, the hydrogen gas, because it has a smaller molar mass, would have a higher average velocity.

202. (D) Graham's law states that the rate of effusion of a gas is inversely proportional to the square root of its molar mass. In comparing two gases, it would be $r_B/r_A = (M_A/M_B)^{1/2}$. Assume that gas A is methane (molar mass of 16 g/mol) and has a relative rate of effusion of 1. The unknown would have an effusion rate of 0.5. Substituting those values into the equation gives: $0.5/1 = (16 \text{ g/mol}/M_B)^{1/2}$. Solving for M_B gives 64 g/mol. Molar mass of oxygen is 32 g/mol, helium is 4 g/mol, carbon dioxide is 44 g/mol, and sulfur dioxide is 64 g/mol. The best choice is sulfur dioxide.

203. (C) The least ideal gas has the largest van der Waals constants. Sulfur dioxide has the largest values.

204. (B) At high pressure there are many gas molecules colliding with the inside walls of the container. This means there are a large number of molecules, which makes the volume of the molecules themselves no longer negligible.

205. (A) Answers C and D are valid at low temperatures. Answer B, concerning the volume of the gas particles being negligible, is related to the pressure. Answer A, the interaction between particles, becomes more significant when the kinetic energies of the particles are lower (at lower temperatures).

206. (C) Reducing the pressure by one-half would cause the volume to double. Doubling the absolute temperature would also cause the volume to double, making the two volumes the same. (Answers B and D are the same answer; therefore, neither can be correct.)

207. (A) The ideal gas equation assumes ideal behavior on the part of the gas. With water vapor, there is a strong interaction between the polar gas molecules, hydrogen bonding, which causes the deviation in pressure from that predicted by the ideal gas equation.

208. **(C)** If the sample weighs 56 grams, then there must be 2 moles of nitrogen present. 1 mole of any ideal gas at STP occupies 22.4 L, so 2 moles would occupy 44.8 L at STP.

209. **(C)** Dalton's law says that the sum of the partial pressures of gases in a gas mixture equals the total pressure. 63 torr + 25 torr + P_{Ar} = 176 torr. The partial pressure of Ar is 88 torr. The percentage of Ar would be (88 torr/176 torr) × 100% = 50%.

210. **(D)** Graham's law states that the rate of effusion of a gas is inversely proportional to the square root of its molar mass. Rounding is very important. In comparing two gases, it would be $r_B/r_A = (M_A/M_B)^{1/2}$. For answer A, Kr and He, it would be [(80 g/mol) / (4.0 g/mol)]$^{1/2}$ > 2 (incorrect); for answer B, O_2 and N_2, it would be [(32 g/mol) / (28 g/mol)]$^{1/2} \approx$ 1 (incorrect); for answer C, H_2 and Xe, it would be [(2 g/mol) / (130 g/mol)]$^{1/2}$ < 2 (incorrect); for answer D, it would be [(80 g/mol) / (20 g/mol)]$^{1/2}$ = 2.0 (correct).

211. **(A)** The latex balloon contains pores that allow gases to effuse through the latex. Since the concentration (and pressure) of gases inside the balloon is greater than that outside the balloon, effusion will take place from inside to outside. Graham's law states that the lower the molar mass, the faster the rate of effusion, so the helium will effuse out faster than the heavier gases that comprise the air. After some helium has effused out, the balloon will not rise as high as before.

212. **(A)** This is a Graham's law problem. The lower the molecular mass (weight), the farther it will travel. Since HX1 traveled the least distance, it has the highest molecular weight.

213. **(D)** Kinetic molecular theory assumes that the intermolecular forces between gas particles are negligible (eliminating answer A), that the average kinetic energies are all the same (eliminating answer B), and that the volumes of the gas particles are negligible (eliminating answer C). It does not assume that the average velocities are the same.

214. **(C)** The sum of the mole fractions must equal 1, so this eliminates answers A and B. The mole fraction is independent of the volume and temperature, so this eliminates answer D. The mole fractions would be the individual gas pressure divided by the total pressure, which is 1.0 atm. So for He, it would be 0.50 atm/1.0 atm = 0.5; Ne: 0.20 atm/1.0 atm = 0.20; and Ar: 0.30 atm/1.0 atm = 0.30.

215. **(C)** You can use the ideal gas equation, PV = nRT, and solve for V (with rounding): V = nRT/P = [(2.0 mol)(0.08206 L·atm/K·mol)(522 K)] / 2.0 atm = 42.8 L. Using 0.1 instead of 0.08206 for the ideal gas constant gives a rounded value of 52.2 L, which because of the rounding should be higher than the unrounded value; 42.8 L is the closest to 52.2 L.

216. **(C)** Use the ideal gas equation and solve for n (with rounding): n = PV/RT = (3.0 atm)(0.080 L) / (0.08 L·atm/K·mol)(300 K) $\approx 10^{-2}$ moles. Therefore, 1.2 g/10^{-2} mol = 120 g/mol. The molar masses are CCl_4 = 154 g/mol, C_8H_{18} = 114 g/mol, C_4H_{10} = 58 g/mol, and CH_3OH = 32 g/mol. C_8H_{18} will give a molar mass closest to the calculated value.

217. **(A)** Whenever a gas is collected over water, there will be water vapor mixed in with the gas. This increases the total pressure of the gas collected.

Chapter 6: Solutions

218. (B) A solution is a homogeneous mixture and will have the same concentration throughout. The solution will have the same freezing point whether the whole solution is tested or only a portion of the solution.

219. (A) Deviation from the predicted van't Hoff factor occurs when solute particles are close together and they can interact, forming ion pairs. The least variance can be expected when the solution is very dilute.

220. (D) In a 0.1 m solution of sodium nitrate, there will be some ion pairing, making the actual van't Hoff factor less than the ideal one. This means that the amount of freezing point depression would be less than expected and the actual freezing point should be higher than the theoretical freezing point that was expected.

221. (A) The use of colligative property measurements (FP depression, BP elevation, and osmotic pressure) is an effective way of measuring the van't Hoff factor, since it appears in their equations. However, density measurements do not involve the van't Hoff factor.

222. (D) The formula of potassium phosphate is K_3PO_4. It is a strong electrolyte, so that 4 moles of ions are formed from 1 mole of the salt. The solution is 1 molal or 1 mole of salt per kg solvent. 1 mole of salt will yield 4 moles of ions, so the solution is 4 mol in particles. The boiling point elevation can be calculated by (0.5°C/mol) × 4 mol = 2°C. That is the amount of elevation; the actual boiling point would be 100°C + 2°C = 102°C.

223. (B) The freezing point is affected by the number of solute particles present. 1 mole of sodium nitrate yields 2 moles of ions, while 1 mole of calcium nitrate yields 3 moles of ions. Therefore, the calcium nitrate solution will exhibit more freezing point depression and will freeze at a lower temperature.

224. (C) In the KCl solution, the interactions that would reduce the van't Hoff factor are between +1 cations and −1 anions. In the copper(II) sulfate solution, the interactions are between +2 cations and −2 anions. Because they have greater charges, these interactions are much stronger.

225. (B) When 1 mole of ammonium sulfate dissociates, it forms 3 moles of ions. Therefore, the van't Hoff factor is 3.

226. (A) When a gas is collected over water, there will be some water vapor mixed with the gas. This comes from evaporation of the water and establishment of the water's vapor pressure.

227. (A) HCN is a weak acid (weak electrolyte), while NaCN is a strong electrolyte and will completely dissociate. If you add an equal number of moles of each to water, there will be more ions in the sodium cyanide solution than in the hydrocyanic acid solution. Since the amount of freezing point depression depends on the number of solute particles, you will have a greater depression of the freezing point with the sodium cyanide solution. Answers C and D involve factors that have nothing to do with the freezing point depression.

228. (A) A solution containing a nonvolatile solute always has a vapor pressure lower than that of the solvent because some of the solute particles will be at the surface, reducing the number of particles that can escape as vapor.

229. (A) Sulfur trioxide is an acidic oxide; therefore, it will react with a base. For this reason, SO_3 is very soluble in a solution of NaOH. The gas reacts with the NaOH as shown by the following reaction: $SO_3(g) + 2 NaOH(aq) \rightarrow Na_2SO_4(aq) + H_2O(l)$.

230. (D) By the solubility rules, answers A, B, and C would be very sparingly soluble, while calcium acetate should be very soluble.

231. (A) The vapor pressure of water is affected by the temperature (the higher the temperature, the higher the vapor pressure), by hydrogen bonding (a determining factor in determining vapor pressure for water), and by the presence of solutes (solutions of nonvolatile solutes have a lower vapor pressure than the pure solvent). The external pressure does not affect the vapor pressure.

232. (B) When the ammonia dissolves in the water, it partially ionizes through the K_b equilibrium of ammonia (eliminating answer A), releasing hydroxide ions, which increases the pH (eliminating answer C). Since a solution has been formed, the boiling point rises. (The boiling point of a solution is always higher than that of the pure solvent unless there is a very volatile solute present.)

233. (A) Decanoic acid is a carboxylic acid that will react with a base. The anion formed will have a hydrophilic end, the carboxylate end.

234. (D) The solution that boils at the highest temperature will be the one with the greatest number of solute particles. Nitric acid is a strong acid, so a 1.0 M nitric acid would contain 2.0 M in ions. The same is true of the 1.0 M NaCl. Ethyl alcohol is a nonelectrolyte, so it would form a solution 2.0 M in particles. Calcium chloride is a strong electrolyte and will yield a solution 3.0 M in ions.

235. (D) The freezing point depression is a result of the lowering of the vapor pressure of the solution.

236. (C) Osmosis and osmotic pressure are colligative properties and are dependent on the number of solute particles present. In answer A, the NaCl is a strong electrolyte, but the acetic acid is a weak electrolyte. Therefore, the solutions will have different numbers of particles and different osmotic pressures. If a semipermeable membrane separates two solutions having the same osmotic pressures, the solvent can pass through in both directions (eliminating answer B). Osmotic pressure is the pressure needed to just stop osmosis (eliminating answer D). In osmosis, the solvent always passes from the solution of lower concentration to one of greater concentration.

237. (B) Magnesium and calcium ions, because of their 2+ charge, would have a greater affinity for the resin and would exchange well. Li^+ would be attracted well to the resin because its small size gives it a higher charge density than sodium ions have. Potassium ions are larger than sodium ions with the same 1+ charge, so their charge density is less, and therefore they have less affinity for the resin than sodium ions do.

238. (A) The sodium, calcium, and aluminum ions are all very stable and difficult to reduce, but Hg^{2+} can be reduced to Hg_2^{2+} or Hg.

239. (D) Compounds A, B, and C are all covalent molecules and should be somewhat volatile. However, $HgCl_2$ is ionic and will not have an appreciable vapor pressure in solution.

240. (B) Adding a low volatility solute will decrease the vapor pressure (vapor pressure lowering is a colligative property). This eliminates answer A. The amount of lowering will be relatively small, not a large amount as in answers C and D. To do the actual calculation, it is necessary to recall that a liter of water is about 55 moles of water; therefore, the mole fraction of water is $55/60 \approx 0.9$, and the vapor pressure of the water will be 0.9×0.30 atm ≈ 0.27 atm.

241. (A) This is a freezing point depression problem. The solution with the greatest number of particles will have the lowest freezing point. Sodium sulfide (answer D) yields 3 particles, sodium phosphate (answer C) yields 4 particles, sodium nitrate (answer B) yields 2 particles, and ethanol (answer A) does not ionize; therefore, it yields 1 particle. The ethanol solution will freeze first (the least particles).

242. (A) The precipitate $Ba_3(PO_4)_2$ will have the strongest ionic bond because of the 3– charge on the phosphate ion giving a greater attractive force to the barium cations. The carbonate and oxalate ions have a −2 charge, and the iodate ion has a −1 charge. Therefore, the Na_3PO_4 solution (answer A) is correct.

243. (A) The reaction would be $NaOH(aq) + HNO_3(aq) \rightarrow NaNO_3(aq) + H_2O(l)$. There are equal numbers of moles of OH^- and H^+, so the only thing left in the solution will be water and $NaNO_3$. Since the concentrations were both 4.00 M and equal volumes were mixed (so the final volume is doubled), the concentration of the $NaNO_3$ solution would be 2.00 M. Since the solution is not as concentrated, it would not freeze at as low a temperature as either of the original solutions (due to colligative properties).

244. (D) A more useful representation of the original formula would be $[Pt(NH_3)_4]Cl_2$, which dissociates upon dissolving to $[Pt(NH_3)_4]^{2+}$ plus 2 Cl^-, for a total of three ions. The $[Pt(NH_3)_4]^{2+}$ ion, like most four-coordinate platinum(II) complexes, is square planar. Answer A cannot be correct, because there are too many ligands present. Answer B, while also four coordinate, does not normally occur for platinum. If the compound were octahedral, there would be no ions released upon dissolution, which eliminates answer C.

245. (A) An acid will react with a basic salt such as sodium sulfide. Neutral salts like calcium nitrate, calcium iodide, and potassium bromide are no more soluble in acid than in pure water. The sulfide ion reacts with hydrogen ions to form hydrogen sulfide gas. As the gas escapes, more sodium sulfide will dissolve to replace the sulfide ion lost by the reaction.

246. (A) The solubility in water is related to having a polar alcohol (all do), but the one with the most polar groups will be the most soluble (answer A, which has 4 alcohol groups).

247. (B) Vapor pressure, freezing point, and boiling point are all colligative properties; that is, they depend on the number of particles present. Density does not depend on the amount present: a big piece of gold has the same density as a small piece of gold.

248. (C) Solubility is affected by the nature of the solute and the solvent: "Like dissolves like." This means that water, a polar solvent, will tend to dissolve polar solvents and not nonpolar ones. Oxygen is nonpolar, and the others are polar.

249. (B) An ideal solution is one that obeys Raoult's law. The more dilute the solution is in terms of particles, the more ideal. Both answers A and C would yield 0.03 M in particles, because you have 3 moles of ions per 1 mole of salt. Answer D would yield a solution that is 0.05 M in ions, but the NaCl solution would yield a solution only 0.02 M in ions.

250. (A) Notice that answers A and B conflict; therefore, one of the two must be the correct answer. There is no need to waste time with the other two answers. Since the vapor pressure of liquid A is higher than that of liquid B, it will take more energy to raise liquid A to the point where its vapor pressure exceeds atmospheric pressure, and thus its boiling point is higher.

251. (A) The pH will not have a significant effect on the gas exchange, and neither will volume changes in the lungs. The temperature will affect both equally, so it will not have an effect. The concentration gradient will have an effect, as the passage of the gases will be from an area of higher concentration to one of lower concentration.

252. (C) The boiling point is a colligative property. The greater the concentration of solute particles, the higher the boiling point. Answers A, B, and C are all soluble salts, while D is a nonelectrolyte. A 2.0 m solution of ethyl alcohol (answer D) will be 2.0 m in particles; 1.0 m sodium nitrate will be 2.0 m in ions; 1.0 m calcium chloride will be 3.0 m in ions; and a 1.0 m solution of aluminum nitrate will be 4.0 m in ions. Since the aluminum nitrate solution has the highest concentration of ions, it will have the highest boiling point.

253. (B) Because all three are covalently bonded, they should have enough vapor pressure for fractional distillation to be effective. All other factors being equal, the first compound to be isolated is the one with the lowest molar mass. AsF_3 has a molar mass of 132 g/mol, PF_3 is 88 g/mol, and SbF_3 is 179 g/mol, so PF_3 would be the first compound isolated.

254. (A) Since a number of precipitates formed, you should look for a solution that contains an ion that forms insoluble substances with the solutions provided. This is true only of $AgNO_3$. It forms white AgCl, yellow AgI, and black AgS.

255. (C) The transition metal carbonates will react with an acid to form carbon dioxide and water. Answer C is nitric acid, a strong acid.

Chapter 7: Kinetics

256. (B) In general, the rate of a reaction doubles for every 10°C rise in temperature. Here there is a 30°C increase; therefore, there are three 10° increases. This leads to a rate increase of $2^3 = 8$.

257. (B) Comparing Experiments 1 and 2, [B] is constant, [A] doubles, and the rate doubles. The reaction is first order in [A]. Comparing Experiments 2 and 3, the [A] is constant, the [B] doubles, and the rate doubles. The reaction is first order also in [B]. The rate law is Rate = k[A][B].

258. (D) If [A] is halved, the rate will decrease to one-fourth $[(1/2)^2 = 1/4]$; if [B] is doubled, the rate will double $(2^1 = 2)$. If both happen at the same time, the rate will be halved $(1/4 \times 2) = 1/2$.

259. (A) In looking at Experiments 1 and 2, the [I⁻] was doubled (other concentrations remaining constant) and the rate doubled. Therefore, the reaction is first order with respect to the iodide ion. Comparing Experiments 3 and 1, the [ClO⁻] was doubled and the rate doubled; therefore, the reaction is first order in hypochlorite. In comparing Experiments 1 and 5, the [OH⁻] decreases by one-half and the rate doubles, meaning that the reaction is −1 with respect to the hydroxide ion.

260. (B) Increasing the hydroxide ion concentration causes the rate to decrease. Substances that do this are inhibitors.

261. (D) Rate = $k[CH_3CHO]^2$; substituting gives 1.6×10^{-2} M s⁻¹ = $k[0.40]^2$; k = $(1.6 \times 10^{-2}$ M s⁻¹$) / (0.40$ M$)^2$. So k = 1.0×10^{-1} M⁻¹ s⁻¹.

262. (A) Comparing Experiments 1 and 3, the oxygen concentration is constant, the NO concentration doubles, and the rate increases fourfold. That means the rate is dependent on $[NO]^2$. Comparing Experiments 3 and 2, where the NO concentration is constant, the $[O_2]$ doubles and the rate doubles. The reaction is first order in $[O_2]$. Therefore, the rate law is R = $k[NO]^2[O_2]$.

263. (A) For a reaction to be slower, the activation energy for the reaction must be higher. It is the magnitude of the activation energy that determines the speed of reaction. The thermodynamic values, ΔH, are irrelevant in a kinetics problem.

264. (B) The rate constant is dependent on the activation energy: the larger the activation energy, the slower the reaction. The temperature also has an effect; increasing the temperature normally increases the rate constant. A catalyst lowers the activation energy and thus affects the rate constant. However, the enthalpy change of the reaction is not related to the rate constant.

265. (A) The acids in answers B, C, and D are all weak acids, and the concentration of the hydrogen ion will be well below 0.1 M. HCl, however, is a strong acid, so the hydrogen ion concentration will be 0.1 M. This is the highest concentration of hydrogen ion, so the HCl solution will cause the release of hydrogen sulfide the fastest.

266. (D) The mechanism consists of three steps; therefore, there must be a peak for each step. This eliminates choices A and B. The central peak corresponds to the second step. The second step is rate determining; therefore, its peak will be the highest relative peak.

267. (D) Normally, changing the concentration of the reactants or products does not change the value of the rate constant, but lowering the activation energy, as with a catalyst, will. The problems states the reaction speeds up as Mn^{2+}(aq) forms; therefore, Mn^{2+}(aq) must be a catalyst. Removing the $MnSO_4$(aq) is removing a catalyst.

268. (A) Answer B would be true for most mechanisms; answer C would be true if the rate-determining step involved only Cl_2; and the temperature dependence is not related to a particular mechanism. The rate-determining step involves the chlorine free radical. If a free radical scavenger inhibits the reaction, then the reaction must involve free radicals.

269. (D) The reaction involves breaking the H–O bond in water, so changing the bond strength would affect the rate. Since the reaction is promoted by light, changing the intensity would have a large effect on the rate. Free radicals are formed during the reaction, so adding a substance that removes free radicals would have a large effect. Adding a nonreacting gas would have a minor effect on the rate, since it would affect the number of effective collisions.

270. (C) In profile I, the reactants are at a lower energy than the reactants (exothermic), and in profile II, the products are of higher energy than the reactants (endothermic).

271. (A) In all but answer A, the products are of lower energy than the reactants, indicating an exothermic reaction. In answer A, the products are of higher energy, an endothermic reaction, and energy will have to be constantly supplied to make the reaction proceed.

272. (A) The rate-determining step should have the highest energy in the energy profile. If the rate-determining step is the first step, then the highest peak should be first. This is what appears in answer A.

273. (B) For the endothermic reaction, you would need to put in 145 kJ/mol to get to the energy level of the reactants and then an additional 285 kJ/mol to get to the top of the activation energy hill. This will be 430 kJ/mol total.

274. (A) Since only one oxygen molecule is produced for every two peroxide molecules that react, the rate of appearance of oxygen should be half the rate of disappearance of peroxide or $(2.0 \times 10^{-1}$ M/s$) / 2 = 1.0 \times 10^{-1}$ M/s.

275. (B) Looking at the energy profile, the activation energy for the forward reaction is much greater than that for the reverse reaction; therefore, the reverse reaction should be faster than the forward reaction.

276. (A) The reaction rate is changed by temperature, which eliminates answer D, and by changes in concentration and pressure of reactants, which eliminates answer B. Decreasing the calcium carbonate particle size will increase the surface area and increase the reaction rate. Changing the concentration of a product does not change the rate of reaction. (This is a kinetics problem, not an equilibrium one.)

277. (D) The OH radical is formed in the rate-determining step, but consumed in the next step. This makes it a reaction intermediate.

278. (C) A catalyst increases the reaction rate by lowering the activation energy of a reaction. This corresponds to the height of the peak in the energy profile.

279. (D) The activation energy of the forward reaction is less than the activation energy of the reverse reaction; therefore, the forward reaction is faster than the reverse reaction. However, there is insufficient thermodynamic data to make a prediction (the entropy change is unknown).

280. (B) The activated complex appears at the top of the activation energy "hill" at position B.

281. (D) There is a general rule that says that for a 10°C increase in temperature, the reaction rate doubles. Conversely, if the temperature were reduced 10°C, the rate would decrease by half. Answers B and C can immediately be eliminated, since most biological reactions will slow down if the temperature is decreased.

282. (D) The reaction with the slowest rate will be the one with the highest activation energy. There is no correlation between ΔH and reaction rate. Be very careful of MCAT questions that mix thermodynamics with kinetics. Very few thermodynamic arguments give correct kinetics answers and vice versa.

283. (D) Br is formed in one step and consumed in the next step. That makes it an intermediate.

284. (B) The rate-determining step in a reaction is the slowest step in the reaction mechanism.

285. (A) Comparing Experiments 1 and 2, where [B] and [C] are constant, the [A] doubles and the rate doubles. The reaction is first order in [A]. Comparing Experiments 2 and 3, the concentrations of [A] and [C] are constant, the concentration of [B] doubles, and the rate does not change. The reaction is zero order in [B]. Comparing Experiments 1 and 4, where the [A] and [B] are constant, the [C] triples and the reaction increases ninefold. The reaction is second order in [C]. Therefore, the rate law is Rate = $k[A][B]^0[C]^2$.

286. (D) Answers A and C are the same: a catalyst lowers the activation energy and affects the rate constant. Since both cannot be correct, this eliminates both. Changing the temperature affects the rate constant. Decreasing the temperature decreases the rate constant, and increasing the temperature increases the rate constant.

287. (B) The rate of reaction is dependent on the activation energy: the larger the activation energy, the slower the reaction. The temperature also has an effect: increasing the temperature normally increases the rate of reaction. A catalyst lowers the activation energy and thus affects the reaction rate. However, the enthalpy change of the reaction is not related to the rate of reaction.

288. (A) One way to express the rate of reaction is to express the change in reactant concentration with time, eliminating answers C and D; this form has a minus sign and is multiplied by 1/a, where a is the coefficient in the balanced equation, eliminating answer B.

289. (C) There is a change in the oxidation state of chlorine, from 0 to −1; the reaction will be faster at elevated temperatures; and both a strong (HCl) and a weak acid (HClO) are formed. The equilibrium constant is not affected by changes in concentrations/pressures. (Remember that equilibrium studies are *separate* from kinetic studies.)

290. (B) The rate law should involve the rate-determining (slowest) step. This eliminates answer A. Answer D is an equilibrium constant, so it is eliminated. Answer C mixes reactants and products. Answer B has the $[H_2O]$ from the slow step and represents the $[Cl•]$ as a function of the Cl_2 by using the one-half exponent.

291. (B) Answer A, the equilibrium answer, is related to thermodynamics, but this is a kinetics question. (This is like comparing apples to oranges and is a common "trick" on the MCAT.) Answers C and D also have nothing to do with rates of reaction. If the activation energy for reaction (2) is greater than the activation energy for reaction (1), the rate of (2) will be slower than that of (1).

292. (A) The correct energy profile would be one in which there was a small peak (first fast reaction), then a much higher peak (second slow reaction), and finally a smaller peak for the third fast reaction.

293. (A) Two moles of the permanganate will disappear while 5 moles of iodine appear. The rate of appearance of iodine, then, should be 5/2 that of the disappearance of permanganate. Therefore, $(5/2) (2.0 \times 10^{-3}\ M/s) = 5.0 \times 10^{-3}\ M/s$.

294. (C) Since the drug is following first-order kinetics, in 4 hours (1 $t_{1/2}$) the level would be down to 50%; in 8 hours (2 $t_{1/2}$), 25%; in 12 hours (3 $t_{1/2}$), it would be at 12.5%; and at 16 hours (4 $t_{1/2}$), it would be at 6.25%. 12 hours would be the closest to 10%.

Chapter 8: Equilibrium

295. (B) The equilibrium constant would have the form of $P_{SO3}{}^2 / (P_{SO2})^2(P_{O2})$. Substituting the pressures (and ignoring units) gives $(10.0)^2 / (10.0)^2(5) = 0.2$. (Notice that the $(10.0)^2$ in both the numerator and the denominator would cancel—no need to do that calculation.)

296. (A) A catalyst affects the speed at which a reaction occurs, not the final result. The reaction will reach equilibrium faster, but the equilibrium constant will be the same.

297. (B) You know initially that the sulfur dioxide and oxygen are in a 2:1 ratio and that the total pressure is 30 atm. That means that the partial pressure of sulfur dioxide is 20 atm and that of oxygen is 10 atm. At equilibrium, $P_{SO2} + P_{O2} + P_{SO3} = 25$ atm. If you let $x =$ pressure of SO_3, then the partial pressure of SO_2 is $(20 - x)$ and the partial pressure of O_2 is $(10 - 0.5x)$. (There is 1:2 stoichiometry between O_2 and SO_3.) Putting these values into the previous equation, $P_{SO2} + P_{O2} + P_{SO3} = 25$ atm, gives $(20 - x)$ atm $+ (10 - 0.5x)$ atm $+ x$ atm $= 25$ atm. Solving for x gives $-0.5\ x = -5$, or $x = 10.0$ atm.

298. (A) Catalysts increase the rates of reaction, allowing the system to reach equilibrium faster, but do not affect the equilibrium concentrations of the reactants and products.

299. (A) The negative ΔH means that the forward reaction is exothermic. Answer B is false, since increasing the temperature will shift the reaction toward the endothermic side, the left in this case, reducing the ammonia concentration. Answer C is false because changing the temperature will disturb the equilibrium. Answer D is false because increasing the temperature will shift the reaction to the left when there are more moles of gas and thus will increase the pressure. Answer A is correct; the kinetic energy of the molecules will increase at a higher temperature.

300. (D) Notice that answers B and D are exactly opposite, so one of these must be the true statement. Look at the K_{sp} values of the two sparingly soluble salts; AgOCN has a larger K_{sp} than AgCN, and therefore it is more soluble. So answer D is the correct answer. Answer C is false because the presence of a common ion reduces solubility, not increases it. In addition, the ΔG of the AgOCN should be more positive than that for the AgCN, since its solubility is much greater.

301. (A) The form of the K_{sp} would be $K_{sp} = [Ag^+][C_2H_3O_2^-]$. Precipitation will occur at the point when the solubility product constant is reached. You have the K_{sp} value and the $[Ag^+]$, so you can use the solubility product constant and solve for the acetate ion concentration $4 \times 10^{-3} = [1 \times 10^{-3}][C_2H_3O_2^-]$. Therefore, $[C_2H_3O_2^-] = 4 \times 10^{-3} / 1 \times 10^{-3} = 4$ M.

302. (C) The equation for the dissolving of the calcium hydroxide is $Ca(OH)_2 \leftrightarrows Ca^{2+} + 2\ OH^-$, so the K_{sp} expression would be $K_{sp} = [Ca^{2+}][OH^-]^2$.

303. (D) The equilibrium equation is $CaSO_4(s) \leftrightarrows Ca^{2+}(aq) + SO_4^{2-}(aq)$. Adding sodium sulfate, an ion common to the equilibrium, forces the equilibrium back to the left, increasing the amount of solid calcium sulfate. Also note that answers A and B are the same; therefore, neither can be correct.

304. (C) Increasing the pressure would cause the equilibrium to shift to the side with the fewest number of moles of gas, the left, decreasing the amounts of CO and H_2 and increasing the amounts of CH_4 and H_2O.

305. (D) Answer A is true; adding a catalyst does not change the value of the equilibrium constant. Answer B is true; adding additional reactant or product to an equilibrium affects the equilibrium amounts and is important. Answer C is true; adding a solid reactant will not affect the equilibrium. Answer D is false; adding a catalyst does not affect the value of an equilibrium constant.

306. (B) Notice that answers A and B are opposite, so one of them must be the correct answer. The oxalate ion is the conjugate base of oxalic acid. This basic ion would be stable only in a basic pH. Therefore, the correct answer is in the high (basic) blood pH. Also note that A and C are the same, which eliminates both answers.

307. (D) The equilibrium is established with Co-60, and then additional Co-59 is added. Even though it is a different isotope of cobalt, chemically it will behave the same and force the equilibrium to the left; however, some of the CoS(s) now contains Co-59, which lowers the fraction of Co-60.

308. (A) According to the solubility rules, the only one of the four choices that is insoluble is barium sulfate.

309. (C) The double arrow indicates an equilibrium state.

310. (A) Increasing the temperature favors the endothermic reaction; therefore, more H_3O^+ is formed, so the pH should slightly decrease.

311. (C) The addition of sulfuric acid will increase the $[H^+]$, which will cause both equilibriums to shift to the left (eliminating answer D as false), lowering the pH (eliminating answer A as false), and since both equilibriums shift to the left, the amount of undissociated H_2S will increase (eliminating answer B as false). There is no reaction to produce H_2 possible with these reactions.

312. (B) If the volume is increased, the equilibrium will shift to the side with the greater number of moles of gas to "fill" the additional space. In this specific equilibrium, there are 2 moles of gas on the left and only 1 mole of gas on the right. Therefore, the reaction will shift to the left.

313. (D) Increasing the temperature favors the endothermic side of the equilibrium, in this case the left side. Increasing the temperature will cause the equilibrium to shift to the left.

314. (A) Air, with its normal concentration of carbon dioxide and oxygen, is taken into the lungs. There, carbon dioxide from the cells is exchanged for oxygen. The exhaled gas is thus lower in oxygen (since some went to the cells) and higher in carbon dioxide (since more was added from the cells).

315. (A) Reducing the volume really is a pressure effect. If the volume is reduced, the pressure increases. Answer C is incorrect since it involves aqueous solutions. Answers A, B, and D involve gases. In the case of answer A, it is a change of 3 moles of gas; in answers B and D, it is a 1 mole change of gas. The greater effect will be in the 3 mole change, answer A.

316. (B) The reaction quotient has the same form as the equilibrium constant. Tripling a concentration that appears in the numerator will triple the reaction quotient.

317. (A) HClO is not a strong acid, so this eliminates answers C and D. The dissociation of $HClO_3$, a strong acid, will reduce the concentration of $HClO_3$, and the equilibrium will shift to the right to replace it.

318. (B) Precipitation will occur if the reaction quotient exceeds the K_{sp}. With rounding, the $[Ca^{2+}]$ would be (10.0 mL × 0.01 M) / 30.0 mL = 3×10^{-3} and the $[F^-]$ would be (20.0 mL × 0.01 M) / 30.0 mL = 7×10^{-3}. The reaction quotient would be $[Ca^{2+}][F^-]^2 = [3 \times 10^{-3}][7 \times 10^{-3}]^2 = 1.5 \times 10^{-7}$. The reaction quotient is less than the K_{sp}, so no precipitation will occur.

319. (C) This is a common ion effect problem. Answer D would not furnish a common ion (CN^-), because the cyanide groups are covalently bonded. The other three answers will furnish cyanide ions, but the calcium cyanide will furnish more cyanide ions than the other two answers and will therefore decrease the solubility the most.

320. (A) Heating the mixture would facilitate the decomposition of the calcium carbonate; adding carbon dioxide would shift the equilibrium to the left; and decreasing the volume would increase the pressure, shifting the equilibrium to the left. Adding CaO, a solid, would have very little effect on the equilibrium.

321. (D) Changing the temperature affects the equilibrium, as does changing the amounts of the gases. Adding a catalyst, however, affects both the forward and reverse reactions equally, resulting in no net change.

322. (B) Changing the temperature affects the equilibrium, as does changing the amounts of the gases, but since there are an equal number of moles on both sides, changing the pressure will have no effect.

323. (B) The equilibrium constant has the form of $K_p = \dfrac{P_{NO}^2}{(P_{N_2})(P_{O_2})}$. Substituting the concentrations gives $K_p = \dfrac{(0.48)^2}{(0.48)(0.24)} = \dfrac{(0.48)}{(0.24)} = 2.0$.

324. (B) This is a common ion problem. Na_2CO_3 is added, and the $[CO_3^{2-}]$ increases, shifting the equilibrium to the left and therefore decreasing the solubility of the $ZnCO_3$, causing more of it to precipitate. The equilibrium constant does not change.

325. (C) The form of the solubility product constant would be $K_{sp} = [Ag^+][CN^-]$. If 1.0×10^{-8} mol/L of AgCN dissolves, the $[Ag^+] = [CN^-] = 1.0 \times 10^{-8}$. Substituting into the K_{sp} expression gives $K_{sp} = [1.0 \times 10^{-8}][1.0 \times 10^{-8}] = 1.0 \times 10^{-16}$.

326. (A) This is a common ion problem. Calcium sulfate is the sparingly soluble salt, so that adding it to a sodium sulfate solution increases the sulfate ion concentration and shifts the equilibrium back to the left, decreasing the solubility of the calcium sulfate.

327. (A) The equilibrium constant expression is a fraction with the product of the right-hand species, each concentration raised to the power of the coefficient in the balanced chemical equation, divided by the product of the left-hand species, each concentration raised to the power of the coefficient in the balanced chemical equation. Therefore, the equilibrium expression would be $K_c = [NO_2]^2 / [NO]^2[O_2]$.

328. (B) Adding CO would force the equilibrium back to the left, decreasing H_2 production. Removing CH_4 would also shift the equilibrium back to the left to replace the methane that was removed, decreasing H_2 production. Increasing the pressure would shift the equilibrium to the side with the fewest number of moles of gas to decrease the pressure: the equilibrium shifts to the left, decreasing H_2 production. Removing H_2 would shift the equilibrium to the right to replace the hydrogen gas, thus increasing H_2 production.

329. (D) The barium oxalate equilibrium is $BaC_2O_4(s) \leftrightarrows Ba^{2+}(aq) + C_2O_4^{2-}(aq)$. Answer A contains a common ion (Ba^{2+}), which will force the equilibrium to the left, meaning lower barium oxalate solubility. Answer B is a possibility, but answer C will give the same result; therefore, neither can be correct. Answer D adds another equilibrium to the mix: $H^+(aq) + C_2O_4^{2-}(aq) \leftrightarrows HC_2O_4^-(aq)$. This equilibrium reduces the oxalate ion, which will cause the solubility equilibrium to shift to the right, increasing the solubility of barium oxalate.

330. (C) The formula of cupric arsenate is $Cu_3(AsO_4)_2$, and the K_{sp} expression is $K_{sp} = [Cu^{2+}]^3[AsO_4^{3-}]^2$ since when the salt dissolves you get 3 cupric ions and 2 arsenate ions. If you let x = moles per liter of the salt that dissolves, then $[Cu^{2+}] = 3x$ and $[AsO_4^{3-}] = 2x$. Substituting into the K_{sp} expression gives $K_{sp} = (3x)^3(2x)^2 = 10^{-37}$. To keep from doing a lot of math, you can approximate $(27x^3)(4x^2) \approx 100x^5$, giving a K_{sp} expression of $10^{-37} = 100x^5$, or $10^{-35} = x^5$. Therefore, $x = 10^{-7}$, $[AsO_4^{3-}] = 2x = 2 \times 10^{-7}$ M.

331. (B) The equilibrium is established with I-131, and then additional I-127 is added. Even though it is a different isotope of iodine, chemically it will react the same and force the equilibrium to the left, reducing the concentration of I-131.

332. (D) Calcium carbonate, $CaCO_3$, is an insoluble salt, so that the concentration of the solution would be very low. The < 0.6 M would be the most logical choice.

Chapter 9: Acids and Bases

333. (B) A buffer has both a weak acid and a weak base component. Answers A, C, and D do not provide a weak base to be with the weak nitrous acid. However, if you add some hydroxide ion, it will react with the nitrous acid to form its conjugate base, the nitrite ion. The moles of NaOH added must be less than the moles of nitrous acid so that the solution will contain both nitrous acid and nitrite ion. Now you have both a weak acid and its conjugate base, a buffer.

334. (B) For a solution to buffer against both an acid and a base, there must be both an acid component and a base component. Sodium carbonate will react only with an acid; ammonium chloride will react only with a base; and calcium chloride will react with neither an acid nor a base. The sodium bicarbonate is amphoteric (amphiprotic) and will react with an acid or a base. It reacts with an acid to form carbonic acid (which decomposes to carbon dioxide and water) and with a base to form the carbonate anion.

335. **(A)** The buffer capacity is the amount of either acid or base that can be neutralized without appreciably changing the pH. The more acid and base is present, the higher the buffer capacity. Therefore, it is advantageous to maximize the concentrations of the conjugate acid-base pair.

336. **(A)** The equilibrium is HOCN $\leftrightarrows$ H$^+$ + OCN$^-$. Adding additional cyanate anion will cause the reaction to shift to the left, reducing the [H$^+$] and increasing the pH. (Remember that increasing the [H$^+$] lowers the pH and vice versa.)

337. **(D)** According to the Henderson-Hasselbalch equation, pH = pK_a + log ([base]/[acid]). In this case, the concentrations of the acid and base are equal, so the pH will equal the pK_a (2×10^{-6}). If the pK_a was 1×10^{-6}, then the pK_a would be 6 and so would the pH. Since the K_a is a little greater, the pK_a and the pH will be between 5 and 6.

338. **(A)** A buffer is composed of a mixture of a weak acid and a weak base. (A conjugate acid-base pair differs by a single H$^+$.) HCl is a strong acid; CaO and Ca(OH)$_2$ are both basic; and C$_2$H$_5$OH$_2$$^+Cl^-$ and C$_2$H$_5$OH are not acid-base pairs. However, HCO$_3$$^-$ and CO$_3$$^{2-}$ are a conjugate acid-base pair and thus will act as a buffer.

339. **(A)** Mixing an equal volume of solutions cuts their concentrations in half, making each solution now 0.25 M. The nitric acid, a strong acid, will react with the ammonia, a weak base, and protonate all of the ammonia to ammonium ions. Left in solution will be ammonium ions and nitrate ions with a concentration of 0.25 M each. The pH will be in the acidic range due to the presence of the weak acid, the ammonium ion.

340. **(D)** Metal oxides in general, when dissolved in water, tend to yield basic solutions due to the formation of hydroxides. However, oxides of metals that have a high oxidation state tend to form acidic solutions. CrO$_3$ is in the highest oxidation state for chromium and forms H$_2$CrO$_4$ (and some H$_2$Cr$_2$O$_7$) when added to water.

341. **(A)** The reaction is HPO$_4$$^{2-}$ $\leftrightarrows$ H$^+$ + PO$_4$$^{3-}$. Therefore, the phosphate ion is the correct choice.

342. **(A)** The hydrogen halides would have the formula HX, where X is any halogen (F, Cl, Br, I, or At). When the hydrogen halide dissociates in water, the anion of the halogen will form. The astinide ion, At$^-$, is the only halide anion in the answers.

343. **(B)** No precipitate will form, since no cation that will form an insoluble salt is added. You can form a gas from a carbonate ion solution by adding an acid, forming H$_2$CO$_3$, which decomposes to carbon dioxide and water.

344. **(A)** CCl$_3$COOH would be the strongest, since it has three chlorines attached to the non-carbonyl carbon. These chlorines are extremely electronegative, pulling electron density from the oxygen atoms and further weakening the O–H bond, making it a stronger acid.

345. **(B)** HNO$_3$ is a strong acid, so that the [H$^+$] = 10^{-2} would give a pH of 2.

346. (A) Every pH unit difference amounts to a power of 10 difference. There is a difference of 6 pH units, so that would be a difference in hydrogen ion concentration of 10^6.

347. (B) Hydrochloric acid is a strong acid. There has been a hundredfold dilution of the volume (10 mL to 1,000 mL). This will reduce the hydrogen ion concentration by two powers of 10 and therefore 2 pH units. In this way, a solution with a pH of 1 would dilute to a pH of 3. However, in this case, a pH increase of 2 would result in a pH of 8, which is clearly impossible for any strong acid solution. The pH must remain below 7.

348. (A) There should be no appreciable reaction with the SiO_2 (the principal component of sand). Calcium carbonate might react, but it would result in a basic solution, since the carbonate ion is a weak base. PbS is insoluble, so no appreciable reaction would be expected. Sulfur dioxide reacts with water to form an acidic solution of sulfurous acid, which may undergo further oxidation to sulfuric acid. This is one of the processes occurring in acid rain.

349. (A) This is a slightly basic pH, so answer B is true. The hydrogen ion concentration is less than the hydroxide ion concentration: pH + pOH = 14, so pOH = 14 − pH, or pOH = 14 − 7.4 = 6.6. So answer C is true. Organic acids would be more soluble in this basic solution than in water, so answer D is true. Calcium carbonate, however, would not be soluble in a basic solution, since the carbonate ion is also a base.

350. (D) Both sulfonic acids and carboxylic acids have acid-base properties that can be used as an indicator. An amine group is basic, so it may also be used in an indicator. An alcohol has no appreciable acid-base properties and thus cannot be used as an acid-base indicator.

351. (A) The reaction is $C_5H_5N(aq) + H_2O(l) \leftrightarrows OH^-(aq) + C_5H_5NH^+(aq)$, and the $K_b = [OH^-][C_5H_5NH^+] / [C_5H_5N]$. If you let $x =$ number of moles per liter of C_5H_5N that hydrolyzes, then $x = [OH^-] = [C_5H_5NH^+]$ and $[C_5H_5N] = 1 - x$. If x is small compared to 1, then $1 - x \approx 1$ and substituting into the K_b gives $1 \times 10^{-10} = [x]^2/[1]$. Therefore, $x = 1 \times 10^{-5} = [OH^-]$. The pOH will be $-\log (1 \times 10^{-5}) = 5$. The pH is $14.000 - pOH = 9$.

352. (C) Answers B and D say the same thing, so neither can be correct. You will be reducing the hydrogen ions but replacing them with potassium ions, so the number of ions remains the same and the conductivity remains constant. Because you are adding a base that reduces the number of hydrogen ions, the pH should increase, not decrease.

353. (C) Nitric acid is a strong acid and had ionized completely, so the $[H^+] = 0.01$ M. The sodium hydroxide reacts with 99% of the initial H^+, but 1% (1/100 or 0.01) of the original hydrogen ion solution remains, so 0.01×0.1 M $= 1 \times 10^{-3}$ is the concentration of the remaining H^+. The pH is therefore 3.

354. (A) Answer A is simply another way of writing phosphoric acid, a polyprotic acid. Acetic acid, B, is a monoprotic acid, and answer D is nitric acid, a monoprotic acid. C is ammonia, a weak base.

355. (D) The production of water from hydrogen and hydroxide ions is exothermic; therefore, the reverse reaction, production of the ions from water, is endothermic. The K_w is based upon this endothermic reaction. Heating a reaction favors the endothermic reactions, so the concentration of hydrogen and hydroxide ions would increase, thereby increasing the value of the K_w.

356. (A) All of the halides except F are conjugate bases of strong acids and will not hydrolyze. However, the fluoride ion is the conjugate base of the weak acid HF. As the conjugate base of a weak acid, it will react with water to produce hydroxide ions, which will make the solution slightly basic: $F^- + H_2O \leftrightarrows HF + OH^-$.

357. (C) There are eight orders of magnitude difference in pH between the two solutions, which is a 10^8 difference in $[H^+]$.

358. (C) Since a strong acid ionizes almost completely, there is a high concentration of products and a very low concentration of reactants, so that the $K_a > 1$.

359. (B) In a polyprotic system, each successive ionization is much smaller than the previous. This eliminates answers A, C, and D. The $[H^+]$ should be the largest, since it is the product of all three ionizations.

360. (B) Since the solution formed was acidic, the $K_a > K_b$.

361. (A) From the table, it is apparent that the smaller the bond energy, the larger the K_a (the stronger the acid). The O, S, Se, and Te are in the same family in the order from top to bottom on the periodic table. The electronegativities are decreasing in this order. Therefore, the acidity (strength) increases as bond energy and electronegativity decrease.

362. (C) All of the hydrogen halides are strong acids with the exception of HF, which is a weak acid. This eliminates answers A, B, and D. In general, the acidity of the hydrogen halides is inversely proportional to the electronegativities of the element. Electronegativities decrease down a family, so acidity should increase.

363. (A) From the K_a values, you can see that HClO is the weaker acid, so the ClO$^-$ is the stronger base. Therefore, it will protonate (accept an H^+) to a greater extent.

364. (D) Answers A, B, and C are bases and would react with the acids, so you could not differentiate between the two. Acetic acid, $HC_2H_3O_2(l)$, is the only acid, and it could act as a solvent.

365. (B) Since amino acids can act as both acids and bases, they are considered amphoteric (amphiprotic).

366. (C) The reaction would be $Ba(OH)_2(aq) + 2\ HNO_3(aq) \rightarrow 2\ H_2O(l) + Ba(NO_3)_2(aq)$. Since it is the reaction between a strong base and a strong acid, it essentially goes to completion. The moles of hydroxide ion equal the moles of hydrogen ion; therefore, there is no excess acid or base. There is no significant hydrolysis of the barium ion or the nitrate ion. So this is a complete neutralization, with a resulting pH of 7.

367. (A) Looking at the reaction, it is obvious that sulfuric acid is donating a hydrogen ion to nitric acid. That makes the nitric acid a base, a hydrogen ion acceptor.

368. (B) The K_{a1} reaction is $H_2C_6H_6O_6(aq) \leftrightarrows H^+(aq) + HC_6H_6O_6^-(aq)$ and $K_{a1} = [H^+]$ $[HC_6H_6O_6^-] / [H_2C_6H_6O_6]$. If you let x = number of moles per liter of $H_2C_6H_6O_6$ that dissociate, then $x = [H^+] = [HC_6H_6O_6^-]$ and $[H_2C_6H_6O_6] = 1 - x$. If x is small compared to 1, then $1 - x \approx 1$, and substituting into the K_{a1} expression gives $1.0 \times 10^{-6} = [x][x] / [1]$. $[H^+] = x = 1.0 \times 10^{-3}$. The additional hydrogen ion from K_{a2} is insignificant because K_{a2} is so small.

369. (C) HNO_3 is a strong acid, so the $[H^+] = 0.010$ or 1×10^{-2}. The pH would then be 2.0.

370. (C) The hydroxide ion is a strong base. All the rest are weak bases.

371. (D) C_2H_5OH is a nonelectrolyte—no base formation will occur. H_3PO_4 is an acid—it will form an acidic solution. NaCl is a strong electrolyte—it will form a neutral salt solution. Na_2CO_3, when it dissolves, will release carbonate ions into the solution. Carbonate ions are weakly basic and will react with the water to form the bicarbonate ion and hydroxide, resulting in a basic solution.

372. (C) Conjugate acid-base pairs differ in a single H^+. All of the answers except answer C fulfill this requirement. In answer C, there is a difference of a hydrogen ion, but the anion portion has changed from nitrate to nitrite.

373. (C) The pK_a is the $-\log K_a$. The second equilibrium is very small and can be ignored, so only K_{a1} is considered. The second equilibrium does not involve H_2S, so it is irrelevant to the question. Therefore, $-\log (K_{a1}) = -\log (1 \times 10^{-7}) = 7$.

374. (C) Ammonium chloride, when dissolved, releases a weak acid, the ammonium ion. This would increase the hydrogen ion concentration (and lower the hydroxide concentration). Neither ethyl alcohol nor sodium chloride would affect the pH of the water. The potassium phosphate, when dissolved, releases the phosphate ion, a weak base. This would react with water to form hydroxide ions, increasing the hydroxide ion concentration.

375. (B) For the titration of a weak acid with a strong base, a weak base will be the predominant species at the equivalence point. An indicator should therefore change color at a slightly basic pH. Thymol blue is the only indicator to change color in a basic range.

376. (C) According to the Henderson-Hasselbalch equation, pH = pK_a + ([base]/[acid]). The $pK_a \approx 12$ (assuming $K_a \approx 1 \times 10^{-12}$), and the ([base]/[acid]) = 1. Therefore, pH = 12 + 0 = 12.

377. (B) The entire discussion of the nitrite ion is meant to distract you as the real exam might. The problem stated that the iron was originally in the +2 state before undergoing oxidation.

378. (A) The strongest base would have the smallest K_a of its conjugate acid.

Chapter 10: Thermodynamics

379. (A) The insulated foam coffee cup greatly reduces the loss of heat from the inside to the outside, that is, from the system to the surroundings. Thus, it is nearly adiabatic.

380. (D) The foam coffee cup will not withstand very high temperatures and cannot be sealed to effectively handle high or low pressures. Therefore, A, B, and C are all true, and the correct answer is D, all of the above.

381. (A) Substituting a weak acid for a strong acid would mean that fewer particles would be consumed because of the weak base equilibrium that would be established. There would be less of an enthalpy change and therefore less of a temperature change.

382. (C) In Experiment 3, the NaOH had to dissolve before the acid-base reaction took place (as in Experiment 2). The dissolving of NaOH is exothermic, resulting in the release of a greater amount of heat and a higher temperature.

383. (A) Since answers A and B are opposite, one of these must be correct. During the reaction, you are going from 3 moles of gas to 2, thereby decreasing the entropy of the system.

384. (D) This is a thermodynamic problem; however, answers A, B, and C refer to kinetics, not thermodynamics. Changing the temperature will affect thermodynamic properties such as $\Delta G°$. It is important to keep the two, kinetics and thermodynamics, separate.

385. (C) The best indicator of whether or not a reaction is spontaneous is the change in the Gibbs free energy, so you can eliminate the first two answers. If a reaction is at equilibrium, then $\Delta G = 0$; this eliminates answer D. If a reaction is spontaneous, its ΔG is negative, and in a nonspontaneous reaction, it is positive.

386. (B) Gases, because of their freedom of motion, have greater entropies than solids. Both $\Delta H_f°$ and $\Delta G_f°$ for any element in its standard state are zero. The standard state for gases (STP) is not the same the thermodynamics standard state. Statement B is true only for a crystalline substance and not for other substances, making the statement false.

387. (A) The $\Delta G_{rxn} = \Delta G°_f$ of the products $- \Delta G°_f$ of the reactants. Both reactants are in their standard state, so all you have to consider is the products. Therefore, $\Delta G_{rxn} = (2 \text{ mol})$ $(-270 \text{ kJ/mol}) = -540 \text{ kJ}$.

388. (C) For a spontaneous process, the enthalpy change is negative and the combustion process has a large positive entropy change. Together, the enthalpy and entropy yield a very negative ΔG. A large negative ΔG gives rise to a large positive K (there are no negative Ks) through $\Delta G = -RT \ln K$.

389. (D) An adiabatic process is one in which there is no net heat exchange between the surroundings and the system.

390. (D) During the reaction, a mole of gas is lost (converted to a solid); therefore, the entropy should decrease.

391. (C) You are adding a solid, $SnCl_2(s)$, and solids, as long as some are present, do not affect the equilibrium.

392. (B) Electrolysis is always nonspontaneous. No hydrogen-hydrogen bonds break. The formation of bonds releases energy.

393. (D) The Gibbs free energy, which is negative for a spontaneous reaction, has the form of $\Delta G = \Delta H - T\Delta S$. If the enthalpy change were negative and the entropy change were negative, then making T sufficiently small would give an overall negative sign to the Gibbs free energy change. Therefore, the reaction could become spontaneous. In all the other choices, a temperature decrease does not lead to a negative ΔG.

394. (C) The change in the internal energy (ΔE) is equal to the heat transferred (q) plus any work done (w): $\Delta E = q + w$. Since heat is being removed, it has a negative sign, and since work is being done on the surroundings, w has a negative sign. Therefore, $\Delta E = -375$ J $- 1,475$ J $= -1,850$ J.

395. (B) Spontaneity is related to both the entropy and enthalpy changes in a system. A negative entropy change can be offset by a large change in the enthalpy. In this case, the liquid loses energy during solidification; this energy loss will increase the entropy of the surroundings. Freezing is spontaneous at any temperature below the freezing point.

396. (A) When ammonia condenses, it must release heat, and therefore, the enthalpy change is negative; as the ammonia goes from the gaseous to the liquid state, the entropy decreases, resulting in a negative sign for the entropy change.

397. (C) When carbon tetrachloride vaporizes, the intermolecular forces must be broken. Carbon tetrachloride is nonpolar, and because of this, the predominant intermolecular force is London dispersion forces. Vaporization requires total breaking of the London dispersion forces to completely separate the molecules from each other. Fusion does not require a complete separation of the molecules, so less energy is necessary.

398. (B) Since ΔG is positive, the forward reaction is nonspontaneous, and so the equilibrium constant, K, will be less than 1.

399. (A) 4.0 g of Ca is 0.10 mole. (Ca = 40 g/mol.) From the reaction, you see that there is a 1:1 stoichiometry between calcium and hydrogen gas, so 0.10 moles of hydrogen gas will be formed. Since the gas is at STP, you know that 1 mol = 22.4 L, so 0.10 mol = 2.24 L. Substituting this volume and pressure into the equation 1 J = 100 L•atm, you get (100) (2.24 L)(1 atm) = 224 J. Since work is being done by the system, the value is negative.

400. (B) The overall enthalpy change would be the difference in the enthalpy of the reactants and products, which is labeled II on the figure.

401. (D) The reaction is endothermic ($\Delta H > 0$) and the entropy is increasing ($\Delta S > 0$). To make this combination ($\Delta H > 0$ and $\Delta S > 0$) spontaneous, it is necessary to increase the temperature.

402. (B) $\Delta G = \Delta H - T\Delta S$. To ensure that ΔG is positive, ΔH must be positive, and ΔS must be negative. If ΔS is negative, then $(-T\Delta S)$ will be positive, making the ΔG positive.

403. (A) Heat (q) is the heat transferred and is related to the internal energy of the system (ΔE) and the work done (w) by the equation $\Delta E = q + w$. This eliminates answers C and D. (Only answer B or C needs to be wrong in order for answer D to be wrong.) Overall, heat is absorbed in the process, making q positive.

404. (A) Answer A should have the greatest increase in entropy because there is a net increase of 6 moles of gas. That greatly increases the entropy of the system. In answers C and D, you are going from gases to gases (not a large increase in entropy), and in answer B, you are going from a liquid to gases (not as large an increase as going from a solid to a gas).

405. (A) Heating water causes an increase in the kinetic energy of the particles. This increase in kinetic energy will increase the entropy of the water molecules because of the increased freedom of motion of the particles. Answer B is incorrect because even at 75°C there is a great deal of hydrogen bonding. Answer C is incorrect because water vaporizes at 100°C, and answer D is incorrect because the viscosity actually decreases.

406. (A) Answer D is incorrect; both compounds are ionic. The formation of Ba(g) yields a large increase in entropy, which makes answer C incorrect. Since the reaction is conducted at high temperatures, the reaction is most probably not spontaneous at room temperature, making answer B incorrect. In the reaction, silicon goes from a 0 charge to a 4+ by losing electrons. This is oxidation.

407. (D) The burning of methane is exothermic, and therefore ΔH is negative, eliminating answers A and C. During the reaction, it goes from 5 moles of gas to 6, increasing the entropy (positive), eliminating answer B.

408. (B) Since there was no heat exchange with the surroundings, q = 0.

409. (A) The ΔG of a reaction at equilibrium equals 0.

410. (B) The enthalpy change is calculated by the (sum of enthalpies of products) − (sum of enthalpies of reactants) = [1 mol (−511 kJ/mole)] − [1 mol (−325 kJ/mol) + 1 mol (0 kJ/mol)] = −186 kJ.

411. (A) Given $\Delta S = -\Delta H/T$, $T = -\Delta H/-\Delta S$. Expressing the enthalpy in J/mol gives (−10,000 J/mol) / (−9.50 J/mol•K). The minus signs cancel, as do the J/mol, leaving T = 1,053 K. If you approximate the 9.5 as 10, you simplify the calculation and get 1,000 K.

Chapter 11: Electrochemistry

412. (D) In an electrolytic cell, you must supply an electrical current to cause a desired redox reaction to occur. Therefore, the cell reaction is nonspontaneous, eliminating answers A and C. By definition, oxidation occurs at the anode, eliminating answer B.

413. (B) Electrolysis of molten magnesium chloride will produce magnesium metal and chlorine gas. The magnesium cations will be reduced (gain of electrons), and the chloride anions will undergo oxidation (loss of electrons). Oxidation is an anode process, and reduction is a cathode process.

414. (B) Reduction occurs at the cathode. The list of half-reactions tells us that reduction will take place with a half-reaction that has a negative voltage. The first one that has a negative voltage is the water decomposition to hydrogen gas.

415. (D) The half-cell reactions are $2 H_2O + 2 e^- \rightarrow H_2 + 2 OH^-$ and $2 H_2O \rightarrow O_2 + 4 H^+ + 4 e^-$. The reaction using hydrogen cations is reduction and occurs at the cathode, and the reaction involving the water forming oxygen is oxidation and occurs at the anode.

416. (B) Since you know that Mg metal is a product of the reaction, then the chlorine half-reaction is reversed and the sign changed on the voltage. Then you can add the reaction along with the half-cell potentials. You get -3.73 volts, which means you must apply a minimum of $+3.73$ volts.

417. (A) The overall cell reaction would be $2 AlCl_3 \rightarrow 2 Al(s) + 3 Cl_2(g)$. Applying reaction stoichiometry (with rounding) gives:

$$\frac{2.7 \text{ g Al}}{1} \times \frac{1 \text{ mol Al}}{27 \text{ Al}} \times \frac{3 \text{ mol Cl}_2}{2 \text{ mol Al}} = 0.15 \text{ mol Cl}_2$$

That gives approximately 0.10 moles of Al and with the 3:2 stoichiometry, 0.15 moles Cl_2.

418. (D) Sodium cannot be produced by electrolysis of an aqueous solution, because the sodium would react violently with water.

419. (D) The reaction is $In^{3+} + 3 e^- \rightarrow In$. You can convert from g In → moles In → moles e^- → coulombs → current. The calculation (with rounding) is:

$$\frac{11.5 \text{ g In}}{1} \times \frac{1 \text{ mol In}}{115 \text{ g}} \times \frac{3 \text{ mol } e^-}{1 \text{ mol In}} \times \frac{100,000 \text{ C}}{1 \text{ mol } e^-} \times \frac{amp \cdot sec}{C} \times \frac{1}{6,000 \text{ sec}} \approx 5.00 \text{ amp}$$

420. (A) The reaction would be $2 F^-(l) \rightarrow F_2(g) + 2 e^-$. The grams of F_2 can be calculated (17 min = 1,029 sec). The calculation (with rounding):

$$\frac{2.0 \text{ amps}}{1} \times \frac{1,000 \text{ sec}}{1} \times \frac{1 \text{ C}}{amp \cdot sec} \times \frac{1 \text{ mole } e^-}{100,000 \text{ C}} \times \frac{1 \text{ mol F}_2}{2 \text{ mol } e^-} \times \frac{40 \text{ g}}{1 \text{ mol F}_2} \approx 0.4 \text{ g F}_2$$

Choose the closest answer, answer A.

421. (B) MgI_2 is a soluble ionic compound releasing ions upon dissolution (electrolyte). It is not easy to oxidize or reduce very stable Mg^{2+}, so it is not likely to be a reducing or oxidizing agent. However, the iodide ion can undergo oxidation, for example, $2 I^- \rightarrow I_2 + 2e^-$, which makes it the reducing agent, the species that facilitates reduction. There are other example of iodide undergoing oxidation (serving as a reducing agent) and no examples where the iodide ion undergoes reduction (serving as an oxidizing agent).

422. (D) Increasing the amount of solid aluminum will not affect the cell potential, eliminating answers A and C. A positive cell potential indicates that the reaction is spontaneous as written. The reverse reaction will therefore be nonspontaneous.

423. (A) For pressure to have an effect on the cell potential, there must be gases involved; this is true only for reactions I and III. Therefore, answer A is correct.

424. (A) Increasing the amount of solid silver will have no effect, because the concentrations of pure solids are not factors.

425. (A) The strongest reducing agent (species being oxidized) is the one that would have the most negative reduction half-cell potential. If you were to flip the reaction around to get the oxidation half-reaction, it would be the one with the greatest oxidation half-cell potential. It would, therefore, be Zn.

426. (C) During the reaction, Ca is undergoing oxidation ($Ca \rightarrow Ca^{2+} + 2 e^-$); that makes it the reducing agent, eliminating answer B. Oxidizing/reducing agents have to be reactant species, eliminating answer D. Cl is undergoing reduction ($Cl_2 + 2 e^- \rightarrow 2 Cl^-$), making it the oxidizing agent.

427. (A) The silver electrode undergoes the following reaction: $Ag \rightarrow Ag^+ + e^-$. This is oxidation and occurs at the anode, eliminating answers C and D. The other electrode must be the cathode and must have a negative charge to draw the silver ions to it in order to undergo reduction to silver metal on the object to be plated.

428. (C) The standard reduction potentials follow the same order within a group as the electronegativities. Fluorine is the most electronegative, and the electronegativities (and standard reduction potentials) decrease going down the group.

429. (A) If the pH increases, the concentration of the hydrogen ion decreases and the cell potential decreases.

430. (A) 9.0 g of Al is approximately 0.33 mole (Al = 27 g/mol). Looking at the half-reaction, there is a 3:1 ratio of moles of electrons to moles of aluminum. Therefore, there will be 1.0 mole of electrons involved in the deposition, and 1.0 mole of electrons = 1.0 faraday.

431. (B) The two balanced half-reactions are $Fe^{2+} \rightarrow Fe^{3+} + e^-$ and $6 e^- + 14 H^+ + Cr_2O_7^{2-} \rightarrow 2 Cr^{3+} + 7 H_2O$. When electron loss and gain between the two half-reactions are equalized (as they must be), it is necessary to multiply the iron half-reaction by 6.

432. (A) To serve only as an oxidizing agent, the compound can never serve as a reducing agent. K_2CrO_4 contains Cr^{6+}, which is the highest oxidation state of chromium. Since there are no higher oxidation states for chromium, it can never serve as a reducing agent. The other three compounds have chromium in an intermediate oxidation state; therefore, any of the three can undergo either oxidation or reduction. (They can behave as either an oxidizing or a reducing agent.)

433. (D) Hexavalent chromium would be chromium in the +6 oxidation state. Assigning oxidations to each of the elements in answers A, B, and C shows that the chromium is in the +6 oxidation state. By process of elimination, answer D has to be the correct answer (as chromium is trivalent).

434. (B) There is no electron exchange in the reaction, so it cannot be a redox reaction, and disproportionation is a type of redox reaction. Chromium is Cr^{6+} throughout. Therefore, the correct answer must be B.

435. (C) If a reaction is spontaneous, its Gibbs free energy change is negative, < 0.

436. (C) This is a disproportionation reaction. In a disproportionation reaction, one substance is both the oxidizing and the reducing agent. Chlorine is oxidized from 0 to +7 and reduced from 0 to −1.

437. (A) In looking at the reaction, aluminum is being oxidized. Therefore, the voltage associated with aluminum is +1.7 V. The overall cell potential is 1.3 V = 1.7 V + x; x, the half-cell potential associated with the indium, is −0.4 V.

438. (C) Platinum is normally inert, resisting redox reaction and acid-base reactions. In order to create a cell, you must have a physical conducting surface on which the electrons can flow. It has to be something upon which one can hook a wire.

439. (A) Oxidation is the loss of electrons and/or numerical increase in oxidation number. In answer B, oxygen is going from the 0 to −2 oxidation state; in answer C, it remains −2; and in answer D, it goes from −1/2 to −1. In answer A, however, it goes from −1 to 0, an increase in oxidation number.

440. (C) In answer A, the electronegative fluorine atoms are bonded to the carbon; in answer B, the electronegative oxygen atoms are bonded to the sulfur; and in answer D, there are not 2 electronegative atoms in the structure. In answer C, however, the 2 electronegative atoms are bonded together, making it a good oxidizing agent.

441. (B) If you look at the half-reaction generating the chromium(III) cation, $6 e^- + 14 H^+(aq) + Cr_2O_7^{2-}(aq) \rightarrow 2 Cr^{3+}(aq) + 7 H_2O(l)$, there is 1 electron involved in the oxidation of the ferrous ion to ferric, but there is a coefficient of 6 in the equation. Pure liquids do not appear in the Nernst equation.

442. (C) Looking at the half-cell potentials, fluorine has a greater positive voltage; therefore, fluorine will oxidize chlorine, not the reverse, eliminating answer A. The halogens are undergoing reduction and are not reducing agents, eliminating answer B, and since fluorine has a larger positive reduction half-cell potential, it would be easier to oxidize chlorine than fluorine. The large positive half-cell reduction potentials indicate that both halogens are good oxidizing agents.

443. (C) Reduction is the gain of electrons, eliminating answers B and D, and the oxidation state decreases, eliminating answer A.

444. (A) Reaction 1 involves the loss of electrons and is oxidation. Reaction 2 involves the gain of electrons and is reduction. This eliminates answers B and C. Oxidation occurs at the anode and reduction occurs at the cathode, eliminating answer D.

445. (C) In looking at the reaction, the Sn^{+4} to Sn^{+2} reaction is reduction and the reaction involving the mercury is oxidation. Therefore, you must reverse the mercury half-reaction and change the sign on the half-cell potential to -0.79 V. Adding the two half-cell potentials together to get the overall cell potential gives 0.15 V $+ (-0.79$ V$) = -0.64$ V. Since the cell potential is negative, the reaction must be nonspontaneous.

446. (B) The oxidizing agent will be the reactant that is being reduced in the reaction. The $Fe^{2+}(aq) \rightarrow Fe^{3+}(aq) + e^-$ is oxidation, so $Cr_2O_7^{2-}(aq)$ is being reduced and is the oxidizing agent.

447. (C) Since this will be a galvanic cell, the cell potential must be positive. One of the reactions must be reversed, changing the sign of the half-cell potential. Electron loss and gain will be equalized by the use of appropriate coefficients, but that will not affect the half-cell potentials. The only way that you can get an overall positive voltage is to reverse the second reaction and add the half-cell potentials: $E°_{cell} = +1.65 + 0.58 = +2.23$ V.

448. (A) To oxidize 1 mole of Sn^{+2} to Sn^{+4} requires 2 moles of electrons; 2 moles of electrons are also required for the half-reaction involving the HClO. However, there was enough tin to require 3 moles of HClO, which means 6 moles of electrons were transferred. The ClO_3^- half-reaction involves 6 moles of electrons for every mole of ClO_3^-, so it would require 1 mole of ClO_3^-.

449. (C) At equilibrium, the cell potential will be zero; eliminating answer D. High pH will reduce the hydrogen ion concentrations, which will be in the denominator in the activity quotient, thus making the cell potential larger.

450. (A) The reaction is $Cr^{3+} + 2\ e^- \rightarrow Cr$. The grams of chromium produced can be calculated (with rounding):

$$\frac{2.00 \text{ amps}}{1} \times \frac{3,600 \text{ s}}{1 \text{ hr}} \times \frac{3 \text{ hr}}{1} \times \frac{1 \text{ C}}{1 \text{ amp·sec}} \times \frac{1 \text{ mol } e^-}{100,000 \text{ C}} \times \frac{1 \text{ mol Cr}}{3 \text{ mol } e^-} \times \frac{52 \text{ g Cr}}{1 \text{ mol Cr}} \approx 3.5 \text{ g Cr}$$

$$\frac{2.00}{1} \times \frac{10,000}{1} \times \frac{1}{100,000} \times \frac{1}{3} \times \frac{52 \text{ g Cr}}{1} \approx 3.5 \text{ g}$$

Chapter 12: Final Review

451. (C) Barium is in the same family as calcium, so it behaves like Ca. Calcium is an important element/ion in many biological functions, including bone formation.

452. (A) The molar mass of $CaSO_4 \cdot 2H_2O$ would be the sum of the atomic masses: $[40 + 32 + (4 \times 16.00) + 2(2 (1.0) + 16.00)]$ g/mol $\approx$ 170 g/mol.

453. (C) Answer A: NH_4BrO_4 (bromine in +7) $HBrO_3$ (bromine in +5); answer B: Cl_2O (chlorine in +1) $HClO_2$ (chlorine in +3); answer C: Cl_2O_5 (chlorine in +5) $HClO_3$ (chlorine in +5); answer D: I_2O_3 (iodine in +3) HIO_4 (iodine in +7).

454. (D) The formula of calcium nitrate is $Al(NO_3)_3$. In the 0.50 M $Al(NO_3)_3$ solution, the $[NO_3^-]$ is 1.5 M (3:1 ratio between nitrate ions and calcium nitrate). Then the solution is diluted: $(0.100 \text{ L})(1.5 \text{ M}) = (1.000 \text{ L})(? \text{ M}); M = 0.15$ M.

455. (B) The heat of vaporization depends on the mass and the strength of the intermolecular forces between the molecules. All of the compounds have approximately the same mass, so the determining factor would be the intermolecular forces. The compound in answer C, a hydrocarbon, would have only weak London forces as the intermolecular force. Answer D would have weak dipole-dipole attraction. Answer A has one hydrogen-bonding site, but answer B has two hydrogen-bonding sites, so its attractive force would be the largest, and so would its heat of vaporization.

456. (A) The heat of vaporization depends upon the molar mass (C) and the intermolecular forces (answers B and D—these two are really the same). The heat of vaporization does not depend on the external pressure.

457. (B) A and C are molecular compounds, which normally have low melting points. B and D are ionic compounds, which normally have high melting points. The charges in CaO are greater than those in NaCl (+2 and −2 versus +1 and −1), and higher charges tend to give higher melting points.

458. (D) $CHCl_3$ is covalently bonded—no ions produced. Answer B is a 1:2 salt, so it would produce 0.90 moles of ions; answer C is a 1:1 salt, so it would produce 0.90 moles of ions; answer D is a 1:3 salt and will produce 1.0 moles of ions.

459. (A) The IV fluid should be such that its osmotic pressure is close to that of blood plasma. The osmotic pressure of water is too low.

460. (B) In order to have hydrogen bonds, there must be a hydrogen atom attached to an O, N, or F. Answers A and C have –OH groups that can hydrogen bond; answer D has an –NH₂ group, which can hydrogen bond. In answer B, the hydrogen atoms are bonded to the carbon.

461. (A) The lower the activation energy, the faster is the reaction. Answer A has the lowest activation energy.

462. (C) Intermediates do not appear in the overall reaction as either reactants or products—they are both formed and consumed during the reaction. This is true of the chlorine and hydroxyl radicals.

463. (D) Adding $CHCl_3$ will not affect the equilibrium, since this is a covalently bonded molecule. Answer B is incorrect, because reducing the Pt concentration would force the reaction back to the left, reducing the cisplatin concentration. Answer C is incorrect, because reducing the ammonia concentration again shifts the equilibrium to the left. Answer D is correct, because increasing the ammonia concentration shifts the equilibrium to the right, increasing the cisplatin concentration.

464. (B) Increasing the pH increases the number of hydroxide ions that would be available to react with the weakly acidic ammonium ion. This would increase the solubility of the ammonium perchlorate, since it is reacting with the base (higher pH).

465. (A) If the reaction quotient of a solution exceeds the K_{sp}, then a precipitate will form. The system is at equilibrium if the reaction quotient equals the K_{sp}.

466. (C) Adding a catalyst affects both the forward and reverse reactions of the equilibrium system equally, so no net change in the equilibrium constant is observed. However, the system will get to equilibrium faster.

467. (D) This is a case of mixing thermodynamics and kinetics. The activation energy gives information about the kinetics of the reaction, but gives no information about the energy change (thermodynamics).

468. (A) Catalysts do not change the heats of reaction; catalysts affect the kinetics (speed) of reaction.

469. (D) The heat of reaction depends on the difference in the energy levels of the reactants and products. It does not depend on the activation energy or transition state.

470. (A) If you calculate all the oxidation numbers for the elements in the reaction, there are changes that identify both an oxidation and a reduction occurring. (You only need to find one oxidation or reduction, since if one is present, the other must be there.) Respiration is a redox process.

471. (B) In batteries, you have a difference of potential that is related to the half-reactions and the reactant concentrations; as the reaction proceeds, the concentrations decrease until there is no longer a potential difference—the cell is at equilibrium.

472. (A) In order for each electron transfer to be spontaneous, the voltage for each electron transfer must be positive. Starting with the largest voltage, you can form a redox reaction with that one and the one with the next highest voltage, so when that one is reversed and the sign on the voltage is made negative, the result is a positive total voltage. Then do that for the third reaction and the second reaction and so on. The correct sequence goes from the one with the highest reduction potential to the lowest, answer A. Changing the order will yield at least one negative sum (nonspontaneous) result.

473. (A) The theoretical yield is the maximum amount of a product that can be formed. Temperature is not normally considered; the amount of NaCl recovered is the actual yield. The concentration of the excess reactant does not matter, since that reactant is in excess. The maximum amount of reactant that can be formed is determined by the amount of the limiting reactant.

474. (A) Balancing the nuclear equation (making sure that there are the same sum of mass numbers and atomic numbers on both sides) gives:

$$_{92}^{235}U + {_0^1}n \rightarrow {_{55}^{140}}Cs + 3\ {_0^1}n + {_{37}^{93}}Rb$$

Answer A is correct.

475. (A) Gallium, Ga, is in the aluminum family, meaning that it has 3 valence electrons, 2 in the 4s and 1 in the 4p, with a filled 3d. There must be 31 total electrons. Answer A has this configuration.

476. (A) Answer A has (13 p + 14 n) versus (14 p + 13 n), answer B has (11 p + 12 n) versus (12 p + 12 n), answer C has (22 p + 26 n) versus (20 p + 28 n), and answer D has (15 p + 17 n) versus (17 p + 19 n). Answer A contains mirror nuclei.

477. (A) Sodium metal has 11 electrons, which eliminates answers C and D, which show only 10 electrons. Sodium's single valence electron is in the 3s orbital, so answer B is the ground-state configuration for sodium. It would be that 3s electron that would gain energy and become excited to the closest higher energy level, the 3p.

478. (A) As you proceed from left to right in a period, the electrons are added to the same major energy level, but the increasing effective nuclear charge causes the atomic radius to decrease. This eliminates answers C and D. Electronegativity is the attraction an atom has for a bonding pair of electrons, so the smaller the atom, the greater the effective nuclear charge and the greater the electronegativity.

479. (D) HCl would be polar, since it is a linear molecule with the atoms having a difference in electronegativities. CH_3Cl would be polar, since the chlorine end of the molecule would pull electron density toward it. H_2S is a bent molecule, like water, with the sulfur being more electronegative than hydrogen. CF_4, even though the elements have a large difference in electronegativities, is a symmetrical molecule (tetrahedral like methane) and thus is nonpolar.

480. (B) Sp hybridization involves the blending of an s and a p orbital. The two resultant sp hybrid orbitals lie 180° apart. This is a linear orientation.

481. (B) These are all members of the same group, the alkaline earth metals. Valence electrons are more easily lost the farther they are from the nucleus. The atomic radius increases from top to bottom within a group. Therefore, the least reactive should be the group member closest to the top of the periodic table, in this case Be.

482. (A) The melting point of an ionic substance is related to the strength of the attractive force between the ions. Both the rubidium cation and the iodide anion are rather large ions, much larger than any of the other ions in the other choices. They are also monovalent ions, so the charge density on each is small. Therefore, the attractive force holding them together is weaker than the other choices, and the melting point is lower.

483. (C) The partial pressure of a gas may be calculated by multiplying the mole fraction of the gas by the total pressure. The total number of moles of gas is 8 moles, and the mole fraction of hydrogen is 1/8. Therefore, the partial pressure of hydrogen is $1/8 \times 1,000$ torr = 125 torr.

484. (A) The rate of effusion of a gas is inversely related to the square root of its molar mass (Graham's law). The molar mass of methane, CH_4, is 16 g/mol, and that of oxygen gas, O_2, is 32 g/mol. The ratio of methane to oxygen effusion would be the square root of the molar mass of oxygen divided by the molar mass of methane: Rate of effusion of methane to oxygen = $(32 \text{ g/mol} / 16 \text{ g/mol})^{1/2} = 1.4$.

485. (A) The volume of a gas is directly proportional to the Kelvin temperature. The initial temperature would be $(227 + 273)$ K = 500 K, and the final temperature would be $(727 + 273)$ K = 1,000 K. Since the final temperature is larger than the initial, the gas will expand, so you must multiply the initial volume by a fraction greater than 1: $(1,000 \text{ K} / 500 \text{ K}) \times 500 \text{ cm}^3 = 1,000 \text{ cm}^3$.

486. (A) For a real gas to approach ideal behavior, the pressure must be low to minimize the effect of the gas's volume (and intermolecular forces of attraction), and the temperature should be high to minimize the attractive forces between gas particles.

487. (C) One mole of $Ce(NO_3)_3$ will furnish 4 moles of ions, 3 moles of nitrate anions, and 1 mole of cerium cations. If you want 1.0 moles of ions, you will need 0.25 moles of $Ce(NO_3)_3$.

488. (A) Answer B is unlikely because it involves a three-body collision. Answer C is unlikely because there is only 1 NO in the (slow) rate-determining step. Answer D is unlikely because there is no NO in the slow step. Answer A is quite likely because in the slow step, you can substitute $[Br_2][NO]$ for the $[NOBr_2]$, and that would give the observed rate law.

489. (A) By Le Châtelier's principle, if a system is at equilibrium and the volume decreases, the reaction shifts toward the side with fewer moles of gas. The value of the equilibrium constant does not change. Therefore, the reaction will shift to the left.

490. (A) The best choice of a buffer is one whose pK_a is as close to the desired pH as possible. Then you can fine-tune the pH by adjusting the concentrations as shown by the Henderson-Hasselbalch equation: $pH = pK_a + \log([base]/[acid])$. HBrO has a K_a closest to the desired pH.

491. (C) Conjugate acid-base pairs differ by a single H^+. The conjugate bases have one less hydrogen ion. In answer A, the phosphate ion has 2 H^+ ions removed. In answer B, the charge on the NH_2 is incorrect. In answer D, only the conjugate acids are shown.

492. (C) $pK_a + pK_b = 14$; $pK_b = 14 - pKa$. The $pK_a = -\log K_a$, so $pK_b = 14 - (-\log K_a)$, or $14 + pK_a$. Therefore, $pK_b = 14 + (\log 1 \times 10^{-10})$.

493. (A) With a $K_a = 5.1 \times 10^{-4}$, the pK_a should be between 3 and 4 based upon the exponent. 3.3 is the only answer in that range.

494. (A) The entropy change for the reaction $(\Delta S°)$ would be the (entropy of the products) − (entropy of the reactants): $1 \text{ mol } (290 \text{ J/mol•K}) - [1 \text{ mol } (198 \text{ J/mol•K}) + 1 \text{ mol } (223 \text{ J/mol•K})] = -131 \text{ J/K}$.

495. (A) As the reaction proceeds, the amount of aluminum metal decreases, not increases, as indicated by answer C. As the reaction proceeds, the concentration of hydrogen ions will decrease and the pH of the solution will increase (becoming more basic).

496. (C) The boiling point is a function of the external pressure, because a substance boils when its vapor pressure equals the external pressure. Decreasing the external pressure decreases the boiling point.

497. (A) Answer A is the only one that is a free radical (having an unpaired electron).

498. (D) Carbon dioxide, with its two double bonds, is linear (eliminating answers A and C). There would be a $C = O$ bond, so there would be only three groups around the C, making the geometry trigonal planar.

499. (D) During a half-life, the amount you start with reduces by one-half. After the first half-life, you have 50% remaining; second half-life, 25%; third half-life, 12.5%; fourth half-life, 6.25%.

500. (D) Since you only need an approximate value of pK_{a1}, then you can make the approximation that a K_a of $8 \times 10^{-3} \approx 1 \times 10^{-2}$. Therefore, the pK_{a1} should be approximately 2. Answer D, 2.1, is very close to this, much closer than any of the other answers are.